Molecular Mechanisms, Diagnoses, and Treatments of Respiratory Diseases

Molecular Mechanisms, Diagnoses, and Treatments of Respiratory Diseases

Guest Editor

Te-Chun Shen

MDPI

Basel • Beijing • Wuhan • Barcelona • Belgrade • Novi Sad • Cluj • Manchester

Guest Editor
Te-Chun Shen
School of Medicine
China Medical University
Taichung
Taiwan

Editorial Office
MDPI AG
Grosspeteranlage 5
4052 Basel, Switzerland

This is a reprint of the Special Issue, published open access by the journal *Biomedicines* (ISSN 2227-9059), freely accessible at: https://www.mdpi.com/journal/biomedicines/special_issues/NR3B90MPHN.

For citation purposes, cite each article independently as indicated on the article page online and as indicated below:

Lastname, A.A.; Lastname, B.B. Article Title. *Journal Name* **Year**, *Volume Number*, Page Range.

ISBN 978-3-7258-5555-1 (Hbk)
ISBN 978-3-7258-5556-8 (PDF)
https://doi.org/10.3390/books978-3-7258-5556-8

Contents

About the Editor

Te-Chun Shen

Te-Chun Shen is an Attending Physician in the Department of Internal Medicine at China Medical University Hospital, Taichung, Taiwan, and an Assistant Professor in the School of Medicine at China Medical University. He also serves as the Teaching Deputy Superintendent and Chief of the Department of Critical Care Medicine at Chu Shang Show Chwan Hospital, Nantou, Taiwan.

Dr. Shen obtained his PhD in Clinical Medical Science from the Graduate Institute of Clinical Medical Science, China Medical University, in 2018, and his MD from the same institution in 2006. His clinical and research interests focus on internal medicine, pulmonary and critical care medicine, and the application of large-scale medical databases to explore respiratory diseases and comorbidities.

He is an active member of several professional societies, including the Society of Internal Medicine, the Taiwan Society of Pulmonary and Critical Care Medicine, the Taiwan Society of Critical Care Medicine, and the Association of Hyperbaric and Undersea Medicine, Taiwan. Dr. Shen also serves on the editorial boards of several academic journals. He has authored and co-authored numerous peer-reviewed articles in SCI journals across fields of internal medicine, pulmonary and critical care medicine, and public health.

biomedicines

Editorial

Molecular Mechanisms, Diagnoses, and Treatments of Respiratory Diseases

Te-Chun Shen [1,2,3]

1 School of Medicine, China Medical University, Taichung 404, Taiwan; chestshen@gmail.com; Tel.: +886-4-22052121
2 Department of Internal Medicine, China Medical University Hospital, Taichung 404, Taiwan
3 Department of Internal Medicine, Chu Shang Show Chwan Hospital, Nantou 557, Taiwan

The Special Issue "Molecular Mechanisms, Diagnoses, and Treatments of Respiratory Diseases" in the journal *Biomedicines* compiles critical advancements in the understanding of respiratory diseases, focusing on their molecular mechanisms, diagnostic approaches, and therapeutic strategies. Inspired by the profound global impact of Coronavirus Disease 2019 (COVID-19), this Special Issue highlights the urgent need for deeper insights into respiratory diseases to facilitate effective clinical interventions and improve patient outcomes. This collection includes original research articles and comprehensive reviews exploring diverse aspects of respiratory diseases [1–8]. Collectively, these studies contribute to a more holistic understanding of respiratory pathogenesis and pave the way for the development of innovative strategies for early diagnosis and precision treatments. Notably, there has been a remarkable surge in the publication of research over the past one to two years addressing a wide range of respiratory diseases, including COVID-19 [9–14], asthma and chronic obstructive pulmonary disease (COPD) [15–18], interstitial lung disease (ILD) [19–21], lung cancer [22–30], bronchiectasis [31–33], pulmonary infection [34], pulmonary hypertension [35,36], and other respiratory disorders [37–40].

Within the context of COVID-19, several studies in this Special Issue focus on pulmonary fibrosis and other long-term sequelae following SARS-CoV-2 infection. For instance, "Long-Term Radiological Pulmonary Changes in Mechanically Ventilated Patients with Respiratory Failure due to SARS-CoV-2 Infection" examines the radiologic and biochemical changes in intensive care unit (ICU) patients with COVID-19 who received mechanical ventilation [1]. This prospective study from Romania highlights the association of specific biomarkers with pulmonary outcomes, including fibrosis, thus suggesting the need for ongoing radiological monitoring in patients with severe COVID-19. Similarly, "Clinicopathological Outlines of Post-COVID-19 Pulmonary Fibrosis Compared with Idiopathic Pulmonary Fibrosis" offers a comprehensive review of the distinguishing clinical, radiologic, and histological features between post-COVID pulmonary fibrosis and idiopathic pulmonary fibrosis (IPF) [2]. This comparison will aid clinicians and researchers in better identifying patient subgroups suited for anti-fibrotic therapies, ultimately guiding future therapeutic approaches.

Furthermore, this Special Issue includes critical analyses of oxygen therapy in the management of hypoxemic respiratory failure. The article "Monitoring the Efficacy of High-Flow Nasal Cannula Oxygen Therapy in Patients with Acute Hypoxemic Respiratory Failure in the General Respiratory Ward: A Prospective Observational Study" investigates the effectiveness of high-flow nasal cannula (HFNC) oxygen therapy [3]. This study supports the use of HFNC as a viable rescue therapy and introduces electrical impedance tomography as a promising tool for monitoring ventilation distribution in real time. These insights could refine respiratory failure management protocols, particularly in general wards where access to advanced ICU resources may be limited.

In COPD research, the articles cover innovative therapeutic targets and risk assessment strategies. For example, "KCa2 and KCa3.1 Channels in the Airways: A New Therapeutic Target" identifies specific potassium channels in the airways as potential therapeutic targets for muco-obstructive diseases [4]. This study explores the roles of KCa2 and KCa3.1 channels in lung physiology and pathology, indicating that modulators of these channels could provide new treatment options for COPD. Additionally, "Associated Factors of Pneumonia in Individuals with Chronic Obstructive Pulmonary Disease Apart from the Use of Inhaled Corticosteroids" delves into pneumonia risks among COPD patients, extending beyond the known association with inhaled corticosteroids (ICSs) [5]. This research suggests that other factors in COPD patients can contribute to pneumonia risk, thereby guiding more nuanced ICS-based therapeutic approaches.

The section on ILD covers the complex interplay between autoimmune diseases and respiratory conditions. In "Better Safe than Sorry: Rheumatoid Arthritis, Interstitial Lung Disease, and Medication—A Narrative Review", the review examines the dual risks of ILD associated with both rheumatoid arthritis (RA) and its treatments [6]. The article highlights the potential for RA therapies, such as disease-modifying antirheumatic drugs, to induce or exacerbate ILD. This study emphasizes the need for cautious and personalized treatment regimens for RA patients with pulmonary complications, which could help clinicians prevent adverse respiratory outcomes. Another relevant article, "Risk of Acute Myocardial Infarction in Pneumoconiosis: Results from a Retrospective Cohort Study", underscores the connection between pneumoconiosis and cardiovascular risks [7]. This population-based study from Taiwan identifies a higher incidence of acute myocardial infarction among pneumoconiosis patients, advocating for integrated cardiovascular monitoring and preventive strategies in this patient group.

This Special Issue also addresses pediatric respiratory health, focusing on obstructive sleep apnea (OSA) in children. "Diagnosis and Treatment of Sleep Apnea in Children: A Future Perspective Is Needed" highlights the current challenges and limitations in diagnosing and treating pediatric OSA [8]. Despite significant repercussions on children's cardiovascular, metabolic, and neurological health, pediatric OSA remains underdiagnosed and undertreated. This article calls for a re-evaluation of diagnostic criteria, including an updated definition of disease severity and the exploration of potential biomarkers for risk assessment. The authors advocate for a shift towards personalized medicine in the management of pediatric OSA, which would enhance treatment efficacy and address the heterogeneity of disease presentation in this population.

In conclusion, this Special Issue provides a multifaceted exploration of respiratory disease research, from the cellular and molecular underpinnings to clinical applications in diagnosis and treatment. Collectively, these studies highlight emerging therapeutic targets, innovative diagnostic techniques, and comprehensive disease management approaches. Looking forward, advancements in molecular and translational respiratory research will likely continue to drive forward personalized medicine, improving outcomes for patients with diverse respiratory conditions. This Special Issue aims to serve as a valuable resource for healthcare providers, researchers, and policymakers as they work to reduce the burden of respiratory diseases on a global scale. By advancing knowledge and promoting collaboration across disciplines, we hope to contribute to a future where respiratory health is proactively managed, enabling better quality of life and survival outcomes for affected patients.

Funding: This research received no external funding.

Conflicts of Interest: The author declares no conflicts of interest.

References

1. Stoian, M.; Roman, A.; Boeriu, A.; Onișor, D.; Bandila, S.R.; Babă, D.F.; Cocuz, I.; Niculescu, R.; Costan, A.; Laszlo, S.Ș.; et al. Long-Term Radiological Pulmonary Changes in Mechanically Ventilated Patients with Respiratory Failure due to SARS-CoV-2 Infection. *Biomedicines* **2023**, *11*, 2637. [CrossRef] [PubMed]
2. Cîrjaliu, R.E.; Deacu, M.; Gherghișan, I.; Marghescu, A.Ș.; Enciu, M.; Băltățescu, G.I.; Nicolau, A.A.; Tofolean, D.E.; Arghir, O.C.; Fildan, A.P. Clinicopathological Outlines of Post-COVID-19 Pulmonary Fibrosis Compared with Idiopathic Pulmonary Fibrosis. *Biomedicines* **2023**, *11*, 1739. [CrossRef] [PubMed]
3. Zhao, Z.; Chang, M.Y.; Zhang, T.; Gow, C.H. Monitoring the Efficacy of High-Flow Nasal Cannula Oxygen Therapy in Patients with Acute Hypoxemic Respiratory Failure in the General Respiratory Ward: A Prospective Observational Study. *Biomedicines* **2023**, *11*, 3067. [CrossRef]
4. Orfali, R.; AlFaiz, A.; Rahman, M.A.; Lau, L.; Nam, Y.W.; Zhang, M. $K_{Ca}2$ and $K_{Ca}3.1$ Channels in the Airways: A New Therapeutic Target. *Biomedicines* **2023**, *11*, 1780. [CrossRef]
5. Lineros, R.; Fernández-Delgado, L.; Vega-Rioja, A.; Chacón, P.; Doukkali, B.; Monteseirin, J.; Ribas-Pérez, D. Associated Factors of Pneumonia in Individuals with Chronic Obstructive Pulmonary Disease (COPD) Apart from the Use of Inhaled Corticosteroids. *Biomedicines* **2023**, *11*, 1243. [CrossRef] [PubMed]
6. Andronache, I.T.; Şuţa, V.C.; Şuţa, M.; Ciocodei, S.L.; Vladareanu, L.; Nicoara, A.D.; Arghir, O.C. Better Safe than Sorry: Rheumatoid Arthritis, Interstitial Lung Disease, and Medication—A Narrative Review. *Biomedicines* **2023**, *11*, 1755. [CrossRef] [PubMed]
7. Chang, J.H.; Shen, T.C.; Chen, K.W.; Lin, C.L.; Hsu, C.Y.; Wen, Y.R.; Chang, K.C. Risk of Acute Myocardial Infarction in Pneumoconiosis: Results from a Retrospective Cohort Study. *Biomedicines* **2023**, *11*, 897. [CrossRef] [PubMed]
8. Solano-Pérez, E.; Coso, C.; Castillo-García, M.; Romero-Peralta, S.; Lopez-Monzoni, S.; Laviña, E.; Cano-Pumarega, I.; Sánchez-de-la-Torre, M.; García-Río, F.; Mediano, O. Diagnosis and Treatment of Sleep Apnea in Children: A Future Perspective Is Needed. *Biomedicines* **2023**, *11*, 1708. [CrossRef] [PubMed]
9. Chikomba, C.; Dlamini, S.; George, J.A.; Pillay, T. COVID Diagnostics: From Molecules to Omics. *Adv. Exp. Med. Biol.* **2023**, *1412*, 141–158. [PubMed]
10. Marjaneh, M.M.; Challenger, J.D.; Salas, A.; Gómez-Carballa, A.; Sivananthan, A.; Rivero-Calle, I.; Barbeito-Castiñeiras, G.; Foo, C.Y.; Wu, Y.; Liew, F.; et al. Analysis of blood and nasal epithelial transcriptomes to identify mechanisms associated with control of SARS-CoV-2 viral load in the upper respiratory tract. *J. Infect.* **2023**, *87*, 538–550. [CrossRef]
11. Yin, S.; Klaeger, S.; Chea, V.A.; Carulli, I.P.; Rachimi, S.; Black, K.E.; Filbin, M.; Hariri, L.P.; Knipe, R.S.; Padera, R.F.; et al. Integrated Immunopeptidomic and Proteomic Analysis of COVID-19 lung biopsies. *Front. Immunol.* **2023**, *14*, 1269335. [CrossRef] [PubMed]
12. Valencia, I.; Lumpuy-Castillo, J.; Magalhaes, G.; Sánchez-Ferrer, C.F.; Lorenzo, Ó.; Peiró, C. Mechanisms of endothelial activation, hypercoagulation and thrombosis in COVID-19: A link with diabetes mellitus. *Cardiovasc. Diabetol.* **2024**, *23*, 75. [CrossRef] [PubMed]
13. Gheorghita, R.; Soldanescu, I.; Lobiuc, A.; Caliman Sturdza, O.A.; Filip, R.; Constantinescu–Bercu, A.; Dimian, M.; Mangul, S.; Covasa, M. The knowns and unknowns of long COVID-19: From mechanisms to therapeutical approaches. *Front. Immunol.* **2024**, *15*, 1344086. [CrossRef]
14. Astin, R.; Banerjee, A.; Baker, M.R.; Dani, M.; Ford, E.; Hull, J.H.; Lim, P.B.; McNarry, M.; Morten, K.; O'Sullivan, O.; et al. Long COVID: Mechanisms, risk factors and recovery. *Exp. Physiol.* **2023**, *108*, 12–27. [CrossRef] [PubMed]
15. Armeftis, C.; Gratziou, C.; Siafakas, N.; Katsaounou, P.; Pana, Z.D.; Bakakos, P. An update on asthma diagnosis. *J. Asthma* **2023**, *60*, 2104–2110. [CrossRef] [PubMed]
16. Schleich, F.; Bougard, N.; Moermans, C.; Sabbe, M.; Louis, R. Cytokine-targeted therapies for asthma and COPD. *Eur. Respir. Rev.* **2023**, *32*, 220193. [CrossRef] [PubMed]
17. Singla, A.; Reuter, S.; Taube, C.; Peters, M.; Peters, K. The molecular mechanisms of remodeling in asthma, COPD and IPF with a special emphasis on the complex role of Wnt5A. *Inflamm. Res.* **2023**, *72*, 577–588. [CrossRef]
18. Li, C.L.; Liu, S.F. Exploring Molecular Mechanisms and Biomarkers in COPD: An Overview of Current Advancements and Perspectives. *Int. J. Mol. Sci.* **2024**, *25*, 7347. [CrossRef]
19. Tseng, C.C.; Sung, Y.W.; Chen, K.Y.; Wang, P.Y.; Yen, C.Y.; Sung, W.Y.; Wu, C.C.; Ou, T.T.; Tsai, W.C.; Liao, W.T.; et al. The Role of Macrophages in Connective Tissue Disease-Associated Interstitial Lung Disease: Focusing on Molecular Mechanisms and Potential Treatment Strategies. *Int. J. Mol. Sci.* **2023**, *24*, 11995. [CrossRef]

20. Koudstaal, T.; Funke-Chambour, M.; Kreuter, M.; Molyneaux, P.L.; Wijsenbeek, M.S. Pulmonary fibrosis: From pathogenesis to clinical decision-making. *Trends Mol. Med.* **2023**, *29*, 1076–1087. [CrossRef] [PubMed]
21. Ji, J.; Zheng, S.; Liu, Y.; Xie, T.; Zhu, X.; Nie, Y.; Shen, Y.; Han, X. Increased expression of OPN contributes to idiopathic pulmonary fibrosis and indicates a poor prognosis. *J. Transl. Med.* **2023**, *21*, 640. [CrossRef] [PubMed]
22. Micheletti, C.; Dhuli, K.; Donato, K.; Gadler, M.; Benedetti, S.; Guerri, G.; Cristofoli, F.; Generali, D.; Donofrio, C.A.; Cominetti, M.; et al. Omics sciences and precision medicine in lung cancer. *Clin. Ter.* **2023**, *174* (Suppl. S2), 37–45.
23. Najjary, S.; de Koning, W.; Kros, J.M.; Mustafa, D.A.M. Unlocking molecular mechanisms and identifying druggable targets in matched-paired brain metastasis of breast and lung cancers. *Front. Immunol.* **2023**, *14*, 1305644. [CrossRef] [PubMed]
24. Pan, X.; AbdulJabbar, K.; Coelho-Lima, J.; Grapa, A.I.; Zhang, H.; Cheung, A.H.K.; Baena, J.; Karasaki, T.; Wilson, C.R.; Sereno, M.; et al. The artificial intelligence-based model ANORAK improves histopathological grading of lung adenocarcinoma. *Nat. Cancer* **2024**, *5*, 347–363. [CrossRef] [PubMed]
25. Song, G.Q.; Wu, H.M.; Ji, K.J.; He, T.L.; Duan, Y.M.; Zhang, J.W.; Hu, G.Q. The necroptosis signature and molecular mechanism of lung squamous cell carcinoma. *Aging* **2023**, *15*, 12907–12926. [CrossRef]
26. Andrikou, K.; Rossi, T.; Verlicchi, A.; Priano, I.; Cravero, P.; Burgio, M.A.; Crinò, L.; Bandini, S.; Ulivi, P.; Delmonte, A. Circulating Tumour Cells: Detection and Application in Advanced Non-Small Cell Lung Cancer. *Int. J. Mol. Sci.* **2023**, *24*, 16085. [CrossRef] [PubMed]
27. Cheng, Z.; Cui, H.; Wang, Y.; Yang, J.; Lin, C.; Shi, X.; Zou, Y.; Chen, J.; Jia, X.; Su, L. The advance of the third-generation EGFR-TKI in the treatment of non-small cell lung cancer (Review). *Oncol. Rep.* **2024**, *51*, 16. [CrossRef]
28. Pongor, L.S.; Schultz, C.W.; Rinaldi, L.; Wangsa, D.; Redon, C.E.; Takahashi, N.; Fialkoff, G.; Desai, P.; Zhang, Y.; Burkett, S.; et al. Extrachromosomal DNA Amplification Contributes to Small Cell Lung Cancer Heterogeneity and Is Associated with Worse Outcomes. *Cancer Discov.* **2023**, *13*, 928–949. [CrossRef] [PubMed]
29. Xie, M.; Vuko, M.; Rodriguez-Canales, J.; Zimmermann, J.; Schick, M.; O'Brien, C.; Paz-Ares, L.; Goldman, J.W.; Garassino, M.C.; Gay, C.M.; et al. Molecular classification and biomarkers of outcome with immunotherapy in extensive-stage small-cell lung cancer: Analyses of the CASPIAN phase 3 study. *Mol. Cancer* **2024**, *23*, 115. [CrossRef] [PubMed]
30. Soloperto, D.; Gazzini, S.; Cerullo, R. Molecular Mechanisms of Carcinogenesis in Pediatric Airways Tumors. *Int. J. Mol. Sci.* **2023**, *24*, 2195. [CrossRef] [PubMed]
31. Raboso, B.; Pou, C.; Abril, R.; Erro, M.; Sánchez, C.; Manzano, C.; Zamarrón, E.; Suarez-Cuartin, G.; González, J. Bronchiectasis. *Open Respir. Arch.* **2024**, *6*, 100339. [CrossRef]
32. Martins, M.; Keir, H.R.; Chalmers, J.D. Endotypes in bronchiectasis: Moving towards precision medicine—A narrative review. *Pulmonology* **2023**, *29*, 505–517. [CrossRef]
33. Mall, M.A.; Burgel, P.R.; Castellani, C.; Davies, J.C.; Salathe, M.; Taylor-Cousar, J.L. Cystic fibrosis. *Nat. Rev. Dis. Primers* **2024**, *10*, 53. [CrossRef]
34. Tunesi, S.; Zelazny, A.; Awad, Z.; Mougari, F.; Buyck, J.M.; Cambau, E. Antimicrobial susceptibility of Mycobacterium abscessus and treatment of pulmonary and extra-pulmonary infections. *Clin. Microbiol. Infect.* **2024**, *30*, 718–725. [CrossRef] [PubMed]
35. Weatherald, J.; Hemnes, A.R.; Maron, B.A.; Mielniczuk, L.M.; Gerges, C.; Price, L.C.; Hoeper, M.M.; Humbert, M. Phenotypes in pulmonary hypertension. *Eur. Respir. J.* **2024**, *64*, 2301633. [CrossRef] [PubMed]
36. Jiang, Y.; Song, S.; Liu, J.; Zhang, L.; Guo, X.; Lu, J.; Li, L.; Yang, C.; Fu, Q.; Zeng, B. Epigenetic regulation of programmed cell death in hypoxia-induced pulmonary arterial hypertension. *Front. Immunol.* **2023**, *14*, 1206452. [CrossRef] [PubMed]
37. Gonnelli, F.; Hassan, W.; Bonifazi, M.; Pinelli, V.; Bedawi, E.O.; Porcel, J.M.; Rahman, N.M.; Mei, F. Malignant pleural effusion: Current understanding and therapeutic approach. *Respir. Res.* **2024**, *25*, 47. [CrossRef]
38. Pederiva, F.; Rothenberg, S.S.; Hall, N.; Ijsselstijn, H.; Wong, K.K.; von der Thüsen, J.; Ciet, P.; Achiron, R.; Pio d'Adamo, A.; Schnater, J.M. Congenital lung malformations. *Nat. Rev. Dis. Primers* **2023**, *9*, 60. [CrossRef] [PubMed]
39. Afzal, A.; Khawar, M.B.; Habiba, U.; Afzal, H.; Hamid, S.E.; Rafiq, M.; Abbasi, M.H.; Sheikh, N.; Abaidullah, R.; Asif, Z.; et al. Diagnostic and therapeutic value of EVs in lungs diseases and inflammation. *Mol. Biol. Rep.* **2023**, *51*, 26. [CrossRef] [PubMed]
40. Gordon, A.C.; Alipanah-Lechner, N.; Bos, L.D.; Dianti, J.; Diaz, J.V.; Finfer, S.; Fujii, T.; Giamarellos-Bourboulis, E.J.; Goligher, E.C.; Gong, M.N.; et al. From ICU Syndromes to ICU Subphenotypes: Consensus Report and Recommendations for Developing Precision Medicine in the ICU. *Am. J. Respir. Crit. Care Med.* **2024**, *210*, 155–166. [CrossRef]

biomedicines

MDPI

Article

Monitoring the Efficacy of High-Flow Nasal Cannula Oxygen Therapy in Patients with Acute Hypoxemic Respiratory Failure in the General Respiratory Ward: A Prospective Observational Study

Zhanqi Zhao [1,2,3,†], Mei-Yun Chang [4,†], Tingting Zhang [5] and Chien-Hung Gow [4,6,7,*]

1. School of Biomedical Engineering, Guangzhou Medical University, Guangzhou 511436, China; zhanqizhao@gzhmu.edu.cn
2. Department of Critical Care Medicine, Peking Union Medical College Hospital, Chinese Academy of Medical Sciences, Beijing 100730, China
3. Institute of Technical Medicine, Furtwangen University, 78054 Villingen-Schwenningen, Germany
4. Department of Internal Medicine, Far Eastern Memorial Hospital, New Taipei City 22060, Taiwan; changmy30@yahoo.com.tw
5. Department of Biomedical Engineering, College of Medicine, Kyung Hee University, Seoul 02447, Republic of Korea; zttysu@khu.ac.kr
6. Department of Internal Medicine, Changhua Hospital, Ministry of Health and Welfare, Changhua 513007, Taiwan
7. Department of Healthcare Information and Management, Ming-Chuan University, Taoyuan 33348, Taiwan

* Correspondence: gowchien@gmail.com; Tel.: +886-4-8298686-2425

† These authors contributed equally to this work.

Abstract: High-flow nasal cannula (HFNC) is widely used to treat hypoxemic respiratory failure. The effectiveness of HFNC treatment and the methods for monitoring its efficacy in the general ward remain unclear. This prospective observational study enrolled 42 patients who had acute hypoxemic respiratory failure requiring HFNC oxygen therapy in the general adult respiratory ward. The primary outcome was the all-cause in-hospital mortality. Secondary outcomes included the association between initial blood test results and HFNC outcomes. Regional ventilation distributions were monitored in 24 patients using electrical impedance tomography (EIT) after HFNC initiation. Patients with successful HFNC treatment had better in-hospital survival (94%) compared to those with failed HFNC treatment (0%, $p < 0.001$). Neutrophil-to-lymphocyte ratios of $\geq$9 were more common in patients with failed HFNC (70%) compared to those with successful HFNC (52%, $p = 0.070$), and these patients had shorter hospital survival rates after HFNC treatment ($p = 0.046$, Tarone-Ware test). Patients with successful HFNC treatment had a more central ventilation distribution compared to those with failed HFNC treatment ($p < 0.05$). Similarly, patients who survived HFNC treatment had a more central distribution compared to those who did not survive ($p < 0.001$). We concluded that HFNC in the general respiratory ward may be a potential rescue therapy for patients with respiratory failure. EIT can potentially monitor patients receiving HFNC therapy.

Keywords: acute hypoxemic respiratory failure; electrical impedance tomography (EIT); high-flow nasal cannula (HFNC); neutrophil-to-lymphocyte ratio (NRL); regional ventilation distribution

1. Introduction

Patients with lung diseases often experience hypoxemia and respiratory distress in general adult respiratory wards. Oxygen therapy is the first-line treatment for these patients. To mitigate hypoxemia, oxygen can be provided through conventional oxygen therapy, including general nasal cannulas, simple face masks, adjustable aerosol masks, and non-rebreathing masks. Non-invasive or invasive ventilators are required for patients with impending respiratory failure. Clinically, appropriate selection of the oxygen delivery device, fraction, and flow is required according to the condition of individuals and the severity of hypoxemia over time. However, the support provided by conventional oxygen therapy may be insufficient for several patients in respiratory wards. The intolerance and discomfort associated with ventilation may also cause treatment failure. The use of a high-flow nasal cannula (HFNC) can reduce the respiratory rate and work of breathing in post-operative or intensive care unit (ICU) patients [1,2].

Over the last 10 years, interest in using HFNC as a first-line treatment has increased in the adult population [3]. HFNC is widely used to treat hypoxemic respiratory failure. The arterial partial pressure of O_2 and fraction of inspired oxygen (PaO_2/FiO_2) ratio are significantly better with HFNC compared with general oxygen therapy [4,5]. Most studies on the effects of HFNC on respiratory failure have focused on ICU or post-operative patients. The effectiveness of HFNC treatment and the methods for monitoring its efficacy in patients with acute hypoxemic respiratory failure (AHRF) in the general ward remain unclear.

This study aimed to assess the efficacy of HFNC therapy in patients with AHRF. We initially investigated the association between initial blood test results and HFNC outcomes. Subsequently, we monitored regional ventilation distributions using electrical impedance tomography (EIT) after initiating HFNC. EIT is a monitoring instrument that analyzes boundary voltage–current data on the chest wall surface during breathing, enabling clinicians to measure and observe the dynamic ventilation distribution and regional lung perfusion [6]. EIT has previously been used to evaluate the effects of HFNC on infants with bronchiolitis [7]. However, the correlation of EIT patterns in patients receiving HFNC therapy remains to be seen, with limited data available. The information gathered in this study may aid in determining whether clinical factors and EIT data predict HFNC treatment outcomes in patients with AHRF in the general respiratory ward.

2. Materials and Methods

2.1. Ethics Statement and Patients

This prospective observational study was approved by Far Eastern Memorial Hospital (FEMH) in Taiwan (FEMH-107139-F). The patients provided written informed consent to participate in this study. Patients with AHRF admitted in the general respiratory ward from April 2019 to December 2021 were screened. The inclusion criteria were as follows: (1) reported dyspnea without oxygen support, (2) age between 20 and 90 years, (3) meeting the AHRF criteria (SpO_2 < 91% or PaO_2 < 60 mmHg without oxygen support, PaO_2/FiO_2 ratio < 300 with oxygen support), and (4) suitable for HFNC therapy as per the attending physician's discretion. The exclusion criteria included pregnancy, brain injury, epilepsy, myocardial infarction, and missing informed consent (patients who did not agree to participate).

2.2. High-Flow Nasal Cannula Therapy and Measurements

The patients included in this study received HFNC therapy using Precision Flow (Vapotherm, Exeter, UK), with initial flow settings of 40 L/min, temperature of 37 °C, and FiO_2 of 0.5, or AIRVO2 (Fisher & Paykel Healthcare, Auckland, New Zealand), with initial settings of flow of 60 L/min, temperature of 37 °C, and FiO_2 of 0.5. The FiO_2 and flow rate were adjusted to maintain the SpO_2 between 92% and 98%. Patient demographics were also recorded at baseline. The comfort level was evaluated using a 7-item questionnaire with a scale of 0 to 10 (from 0, no discomfort, to 10, totally intolerable). Failure of HFNC treatment was defined as death during HFNC treatment, the necessity of masked bilevel positive airway pressure or endotracheal tube intubation, and mechanical ventilator support to maintain ventilation.

Electrical impedance tomography (EIT) was routinely performed on patients with HFNC unless contraindications for EIT were present (e.g., pacemaker, automatic implantable cardioverter defibrillator, implantable pumps, large wound on the chest) or patients refused to use EIT. A 16-electrode EIT electrode belt was placed on the chest at the fifth intercostal space, with a reference electrode positioned on the abdomen (PulmoVista

500; Dräger Medical, Lübeck, Germany). An alternating current was applied during the sequential rotation. The frequency and amplitude of the current were automatically determined based on the background noise in the measurement environment. The surface potential difference between the adjacent electrode pairs was measured and recorded at a frequency of 20 Hz. EIT was performed with the patients spontaneously breathing in the supine position at three time points: T0, 30 min before the start of HFNC; T1, 2 h after HFNC initiation; and T2, 24 h after HFNC initiation. A 15-min measurement was recorded at each time point. Suction and positional changes were avoided during the EIT measurements. The images were reconstructed using the manufacturer's software (EIT Data Review Tool version 6.3; Dräger Medical, Lübeck, Germany). Custom software programmed in MATLAB R2015 (MathWorks Inc., Natick, MA, USA) was used for the offline analysis of the EIT data.

2.3. Electrical Impedance Tomography Data Analysis

Functional EIT (fEIT) tidal variation (TV) was derived by computing the difference between the end-expiration and end-inspiration images, capturing the variation during tidal breathing. Tidal images of 1 min were averaged to increase the signal-to-noise ratio:

$$TV_i = \frac{1}{N}\sum_{n=1}^{N}\left(\Delta Z_{i,Ins,n} - \Delta Z_{i,Exp,n}\right) \tag{1}$$

where TV_i is the pixel i in the fEIT image, N is the number of breaths within the analyzed period, and $\Delta Z_{i,Ins}$ and $\Delta Z_{i,Exp}$ are the pixel values in the raw EIT image at the end inspiration and end expiration, respectively. When TV_i was <0, a value of 0 was assigned to TV_i.

Three EIT-derived indices were investigated and assessed to quantify the spatial and temporal distributions of ventilation. The global inhomogeneity (GI) index was computed from the tidal EIT images to characterize the variability in ventilation [8]:

$$GI = \sum_{l \in lung}\left|TV_l - Median\left(TV_{lung}\right)\right| / \sum_{l \in lung} TV_l \tag{2}$$

TV represents the differential impedance value in the tidal images, TV_l signifies the pixel within the identified lung area, and pixel l is classified as part of the lung region if TV_l is >10% × max (TV). TV_{lung} encompasses all the pixels depicting the lung region. A heightened GI indicates substantial variation among the impedance values of the tidal pixels. The center of ventilation (CoV) illustrates the distribution of ventilation affected by factors such as gravity or different lung conditions (weighted relative impedance values based on anteroposterior coordinates) [9]:

$$CoV = \sum(y_i \times TV_i) / \sum TV_i \times 100\% \tag{3}$$

where TV_i is the impedance change in the fEIT image for pixel i, y_i is the height of pixel i, and the value is scaled such that the bottom of the image (dorsal) is 100% and the top (ventral) is 0%.

The tidal image was divided into four horizontal and anterior-to-posterior segments of equal height (regions of interest (ROIs)). The ventilation distributions in these regions were calculated and are denoted as ROIs 1–4.

The regional ventilation delay (RVD) index describes the regional delay in ventilation by comparing the rising time of the pixel impedance to the global impedance curve [10], which can be used to assess the tidal recruitment and derecruitment:

$$RVD_l = t_{l,40\%} / T_{inspiration,global} \times 100\% \tag{4}$$

where $t_{l,40\%}$ is the time required for pixel l to reach 40% of its maximum inspiratory impedance change and $T_{inspiration,global}$ denotes the inspiration time calculated from the global impedance curve.

2.4. Statistical Analyses

Data analysis was performed using MATLAB R2015 (MathWorks Inc., Natick, MA, USA). Clinical data were analyzed based on categorical variables, which were compared using the chi-squared test. Fisher's exact test was applied when the expected value was <5. In the hospital ward, survival was calculated using the Kaplan–Meier method. Differences in the survival curves were measured using the log-rank test or Tarone-Ware test. Whether HFNC success or failure was associated with demographics, FiO_2, flow rate, or blood cell count was also explored. The Lilliefors test was used to test for normality. For normally distributed data, results were expressed as mean ± standard deviation. For non-normally distributed data, the results were presented as medians (minimum–maximum). A two-way analysis of variance was used to evaluate the differences between the three time points and HFNC outcomes. $p < 0.05$ was considered statistically significant.

3. Results

Forty-two patients were enrolled in this study, and EIT measurements were available for twenty-four patients. The clinical characteristics of all patients are listed in Table 1.

Table 1. Clinical characteristics of patients receiving HFNC therapy for acute hypoxemic respiratory failure.

Clinical Characteristic	Values
Patients, *n*	42
Age, years	
Median (minimum–maximum)	74.5 (52–88)
Sex, *n* (%)	
M	24 (57)
F	18 (43)
Smoking status, *n* (%)	
Current/Ever	23 (55)
Never	19 (45)
Primary cause of respiratory failure, *n* (%)	
Pneumonia	33 (78)
Obstructive lung diseases	5 (12)
Lung cancer	4 (10)
Heart failure, *n* (%)	
Yes	3 (7)
No	39 (93)
HFNC efficacy, *n* (%)	
Success	32 (76)
Failure	10 (24)
Event (days ± SD)	
Hospital admission	29.4 ± 23.1
HFNC to discharge	20.8 ± 17.8
HFNC use	11.3 ± 8.9
Survival/Death, *n* (%)	
Survival	30 (71)
Death	12 (29)

Abbreviations: F, female; HFNC, high-flow nasal cannula; M, male; *n*, number; SD, standard deviation.

The primary cause of acute hypoxemic respiratory disease was pneumonia, followed by obstructive lung disease and advanced lung cancer. No patient with interstitial lung disease was included in the study subjects. Ten patients failed HFNC treatment (10 of 42, 24%). Among them, an advanced lung cancer patient who suffered from pulmonary lymphangitic carcinomatosis with a do-not-intubate order was treated with HFNC oxygen therapy for four days but failed. Subsequently, she received non-invasive mechanical ventilator support and passed away after one day. Another nine patients who failed HFNC treatment had pneumonia as the primary cause, including one patient who progressed to acute respiratory distress syndrome and received endotracheal intubation, mechanical ventilator support, and ICU admission; five patients received non-invasive ventilation (masked bilevel positive airway pressure) support; two patients had do-not-intubate orders and passed away; and one patient had a hospital-acquired coronavirus disease 2019 (COVID-19) and was under non-rebreathing mask support and eventually died.

The remaining 32 (76%) patients were successfully weaned off HFNC, and their oxygen demand was reduced to a simple mask, nasal cannula, or room air oxygenation after their disease became stable. Twenty-four patients with pneumonia, five patients with obstructive lung diseases, and three patients with advanced lung cancer discontinued HFNC and converted to conventional oxygen therapy after proper treatments. Among the lung cancer cases, two patients with malignant pleural effusion successfully removed HFNC after pigtail catheter drainages and treatment with albumin plus diuretics; one patient who experienced hemoptysis and aspiration was stabilized after medical treatment and weaned off HFNC. The overall in-hospital mortality rate was 29% (12 of 42). Two patients successfully received HFNC treatment during the AHRF episode but died later due to other causes.

The potential prognostic factors associated with HFNC success or failure were analyzed (Table 2). Patients in the HFNC success group had better in-hospital survival (survival rate, 94%, 30 of 32) compared with those in the HFNC failure group (survival rate, 0%, 0 of 10; $p < 0.001$). Other factors were similar between the two groups. Oxygen delivery was also analyzed in the HFNC success and failure groups. The delivered FiO_2 levels were divided into low (<60%) and high (≥60%). Oxygen flow was divided into low (<40 L/min) and high (≥40 L/min). The selected cut-off FiO_2 and flow rates were based on an increased risk of intubation among critical patients treated with HFNC [11]. Patients were divided into low- and high-FiO_2 groups, and flow rates were delivered on Day 1 of HFNC. No significant differences were observed between the success and failure groups. Other cut-off points for FiO_2 (40 and 50%) and flow rates (20 and 30 L/min) were also explored. Similar results were obtained.

Table 2. Factors associated with HFNC therapy treatment outcomes in patients with acute hypoxemic respiratory failure.

Factor	HFNC Success ($n = 32$)	HFNC Failure ($n = 10$)	p-Value
Age, n			
<75	16	5	1.000
≥75	16	5	
Sex, n			
M	18	6	1.000
F	14	4	
Smoking status, n			
Current/Ever	18	5	0.729
Never	14	5	
Survival/Death, n			
Survival	30	0	<0.001 *
Death	2	10	
Primary cause of respiratory failure, n			
Pneumonia	24	9	0.564
Obstructive lung disease	5	0	
Lung cancer	3	1	

Table 2. *Cont.*

Factor	HFNC Success (n = 32)	HFNC Failure (n = 10)	p-Value
Heart failure, n			
Yes	3	0	1.000
No	29	10	
HFNC, FiO_2 on Day 1 (%), n			
<60	21	4	0.268
≥60	11	6	
HFNC, flow on Day 1 (L/min), n			
<40	21	8	0.466
≥40	11	2	
Hemoglobin (mg/dL), n			
≥9	27	5	0.040 *
<9	5	5	
White cell count (per dL), n			
≥12,000	12	4	1.000
<12,000	20	6	
Platelets (per dL), n			
≥80,000	29	9	1.000
<80,000	3	1	
Neutrophil-to-lymphocyte ratio, n			
≥9	11	7	0.070
<9	21	3	
pH level, n			
≤7.34	4	1	1.000
7.35–7.45	17	6	
≥7.46	11	3	
CO_2 level (mmHg), n			
<40	13	4	0.625
=40–55	13	7	
≥55	6	3	
BUN (mg/dL) on Day 1, n			
≤25	15	6	0.719
>25	17	4	
Creatinine (mg/dL) on Day 1, n			
≤1.3	26	7	0.660
>1.3	6	3	
Albumin (mg/dL) on Day 1, n			
<3.0	9	6	0.128
≥3.0	23	4	

Abbreviations: F, female; HFNC, high-flow nasal cannula; M, male; n, number. p-values were calculated using a two-sided chi-squared test. * Statistically significant values (p < 0.05).

Complete blood cell count, differential count, venous blood gas (VBG), and biochemical examination (blood urea nitrogen (BUN), creatinine, and albumin) results were analyzed. We observed that patients in the HFNC treatment failure group had a higher frequency of anemia (Day 1 hemoglobin (Hgb) < 9 mg/dL), with 50% (5 of 10), compared with those in the success group (16%, 5 of 17; p = 0.040). Leukocytosis and platelet count were not prognostic factors. The neutrophil-to-lymphocyte ratio (NLR), which assesses inflammatory or infectious conditions and represents physiological stress, was also evaluated. We categorized NLR levels into two groups, ≥9 and <9, as NLR ≥ 9 has been associated with predicting mortality in critically ill patients with pneumonia [12]. Although a trend was observed, no significant difference was found between the two groups. Other factors, such as pH value (acidosis vs. normal vs. alkalosis), CO_2 level in the VBG, BUN level, creatinine value, and albumin level, were not different between the two groups.

Most patients tolerated HFNC with limited discomfort during body turnover and movement (Table 3). Univariate analyses of prognostic factors for in-hospital survival were performed using the Kaplan–Meier method and the log-rank test or the Tarone-Ware test (Table 4). HFNC treatment failure was associated with poorer survival outcomes for both ward survival ($p < 0.001$, Figure 1a) and survival after HFNC treatment ($p < 0.001$, Figure 1b). Similar to the prognostic factors associated with HFNC success or failure, age, sex, smoking status, and major diseases were not predicting factors for in-hospital survival either for the ward or after HFNC treatment. The delivered FiO_2, flow rate, Hgb level, WBC count, or platelet count could not separate the survival outcomes. Patients with NLRs $\geq$9 were associated with shorter in-hospital survival (39 days) compared with those with NLRs <9 (110 days), but the difference was not significant ($p = 0.062$ using the log-rank test, Figure 1c). Similar results were observed for hospital survival after HFNC treatment (31 vs. 41 days, NLR $\geq$ 9 vs. <9, $p = 0.055$ using the log-rank test and $p = 0.046$ using the Tarone-Ware test, Figure 1d).

Table 3. Seven-item questionnaires to evaluate the comfort level of high-flow nasal cannula oxygen therapy for patients with acute hypoxemic respiratory failure.

Item	Questionnaire	Score ± SD
1	How comfortable is your nose or face when using oxygen equipment?	2.0 ± 1.5
2	How comfortable is your mouth/nose/throat (whether it is dry) when using oxygen equipment?	2.4 ± 1.8
3	How comfortable do you feel to swallow when using oxygen equipment?	2.2 ± 1.6
4	How comfortable do you feel during eating when using oxygen equipment?	2.4 ± 2.0
5	What extent do you feel that the use of oxygen equipment affects coughing?	2.4 ± 1.8
6	How much do you feel about body turnover when using oxygen equipment?	3.8 ± 1.9
7	How much do you feel about movement and activity using oxygen equipment?	3.9 ± 2.0

We recorded scores from Day 1 to Day 3 and expressed them as an average ± standard deviation (SD); the scores were defined as follows: 0 (no discomfort), 3 (slightly uncomfortable but acceptable), 5 (not very comfortable), 7 (uncomfortable and unbearable), and 10 (totally intolerable).

Table 4. Prognostic factors for in-hospital survival of patients with acute hypoxemic respiratory failure under HFNC therapy.

		Ward Survival		HFNC to Discharge	
Factor	**Patients *n***	**Median (Days)**	***p*-Value**	**Median (Days)**	***p*-Value**
Age, years					
<75	21	47	0.705	41	0.473
≥75	21	110		103	
Sex					
M	24	42	0.346	41	0.574
F	18	110		38	
Smoking status					
Current/Ever	23	42	0.599	41	0.850
Never	19	110		38	
HFNC efficacy					
Success	32	23	<0.001	103	<0.001
Failure	10	110		13	
Primary cause of respiratory failure					
Pneumonia	33	47	0.890	38	0.716
Non-pneumonia	9	47		41	

Table 4. *Cont.*

Factor	Patients n	Ward Survival Median (Days)	Ward Survival p-Value	HFNC to Discharge Median (Days)	HFNC to Discharge p-Value
HFNC, FiO_2 on Day 1 (%)					
<60	25	110	0.136	103	0.282
≥60	17	42		38	
HFNC, flow on Day 1 (L/min)					
<40	28	47	0.948	38	0.760
≥40	14	110		103	
Hemoglobin (mg/dL)					
≥9	32	NR	0.356	38	0.380
<9	10	47		41	
White cell count (per dL)					
≥12,000	16	NR	0.973	NR	0.358
<12,000	26	47		41	
Platelets (per dL)					
≥80,000	38	110	0.894	103	0.577
<80,000	4	47		38	
Neutrophil-to-lymphocyte ratio					
≥9	18	39	0.062	31	0.055
<9	24	110		41	0.046 * (T–W)
pH level					
7.35–7.45	23	47	0.928	38	0.776
Abnormal	19	NR		NR	
CO_2 (mmHg) level					
40–55	19	47	0.733	38	0.290
Abnormal	23	110		41	
BUN (mg/dL) on Day 1					
≤25	21	42	0.522	38	0.526
>25	21	110		103	
Creatinine (mg/dL) on Day 1					
≤1.3	33	47	0.357	41	0.540
>1.3	9	29		31	
Albumin (mg/dL) on Day 1					
<3.0	15	47	0.493	38	0.387
≥3.0	27	110		41	

Abbreviations: F, female; HFNC, high-flow nasal cannula; M, male; n, number. p-values were calculated using the Kaplan–Meier method and using the log-rank or Tarone-Ware test (T–W test) to measure all differences in survival. * Statistically significant values ($p < 0.05$).

The EIT data were available for 24 patients, including 18 patients with pneumonia, 4 patients with obstructive lung disease, and 2 patients with advanced lung cancer. Among them, 20 patients were successfully treated with HFNC, and 4 failed. Regarding in-hospital survival, 18 patients remained alive, while 6 patients died. All EIT data retrieved from patients who failed HFNC treatment or died were from those who had pneumonia as the primary cause of AHFC. No significant differences were found in the EIT-based parameters of GI or RVD among the three time points or between the HFNC success and failure groups. At T1 and T2, the ventilation distribution of patients with successful HFNC treatment was more significant towards the center (CoV closer to 50%). The data were significantly different between HFNC success and failure at T0, T1, and T2 ($p < 0.05$, Figure 2a). Similarly, at T1 and T2, patients who survived HFNC treatment had a more central distribution compared with those who did not survive ($p < 0.001$, Figure 2b).

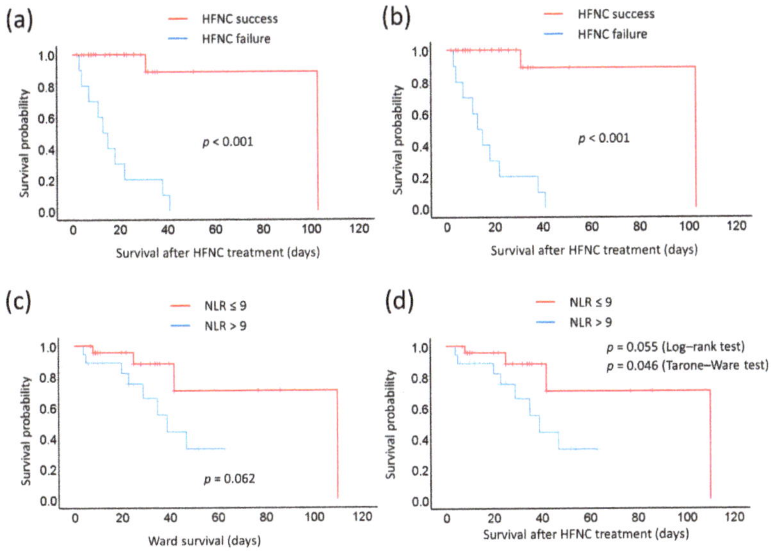

Figure 1. Kaplan–Meier curves of in-hospital survival for acute hypoxemic respiratory failure under HFNC oxygen therapy: (**a**) The 'ward survival' of patients with HFNC success or HFNC failure. (**b**) The 'survival after HFNC treatment' of patients with HFNC success or failure. (**c**) The 'ward survival' of patients with high neutrophil-to-lymphocyte ratios (NLRs) of ≥9 or low NLRs (NLR < 9). (**d**) The 'survival after HFNC treatment' of patients with high NLRs of ≥9 or low NLRs of <9. (n = 42). HFNC, high-flow nasal cannula.

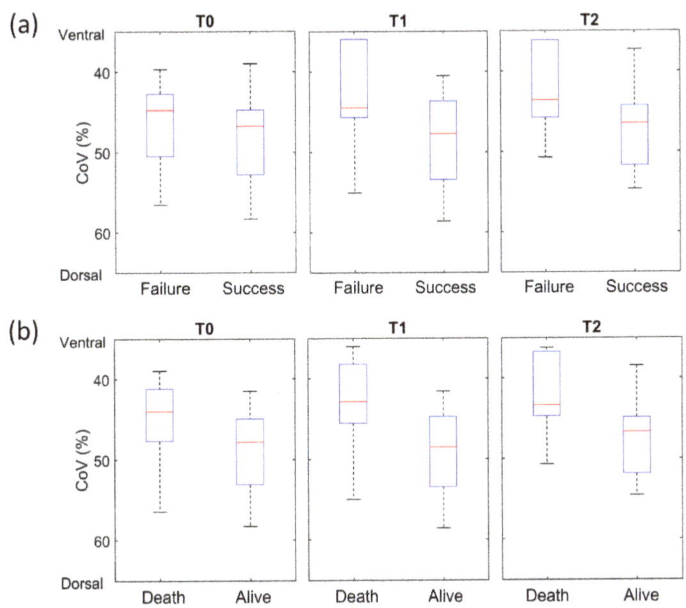

Figure 2. EIT-derived CoV at different times: (**a**) between the HFNC success and failure groups and (**b**) between the survival and death groups. T0, 30 min before HFNC; T1, 2 h after HFNC initiation; and T2, 24 h after HFNC initiation. (n = 24). EIT, electrical impedance tomography; HNFC, high-flow nasal cannula.

4. Discussion

In this prospective observational study, we demonstrated that HFNC therapy was clinically applicable for managing patients with acute hypoxemic respiratory failure in a general respiratory ward. The success rate of weaning off HFNC was 76%. The in-hospital mortality rate was 29%. EIT can potentially help monitor patients receiving HFNC therapy.

HFNC can be beneficial, feasible, and safe for patients with AHRF in the general ward, including those with lung cancer, chronic obstructive pulmonary disease, and mild-to-moderate adult respiratory distress syndrome [13]. Recently, a randomized multicenter clinical trial study in the ICU demonstrated that HFNC treatment could improve AHRF compared with conventional oxygen therapy or non-invasive ventilators and that the 90-day survival rate was better among patients with HFNC compared with those receiving conventional oxygen therapy and non-invasive ventilation [14]. Several studies have described the early predictors of HFNC outcomes in AHRF. Clinical factors, such as baseline heart rate, alveolar–arterial PO_2 difference, Sequential Organ Failure Assessment score, and vasopressor use, are significantly higher in HFNC failure groups than in HFNC success groups among patients with specific diseases admitted to the ICU [15–17]. These results suggest that the initial disease severity and organ dysfunction may be good predictors of HFNC failure in patients with acute respiratory failure. In the present study, we found that NLRs > 9 could possibly predict HFNC treatment failure in the general respiratory ward. The NLR is considered a potential marker for predicting the requirement of a high-flow oxygen nasal cannula and invasive mechanical ventilation in acute hypoxemic respiratory cases [18–20]. A high NLR in early AHRF indicates the severity of the disease and organ dysfunction and has been observed in several conditions associated with tissue damage-induced systemic inflammatory response syndrome [21]. Although several studies and our data have demonstrated that HFNC treatment failure certainly has a higher NLR [18,20,22], the cut-off NLR value to determine HFNC treatment outcomes requires further examination. Other factors, such as pH, CO_2 levels in VBG, BUN, creatinine, and albumin levels, were not different between the two groups.

Patients in the failure group developed anemia more frequently than those in the successful group ($p = 0.040$). Research reporting the relationship between anemia and outcomes of HFNC oxygen therapy in AHRF patients is rare. However, previous studies have shown that anemia patients with hypoxemic respiratory failure have poor treatment outcomes. Anemia patients in the ICU had adverse outcomes and a higher extubation failure rate [23,24]. Patients who survived acute respiratory distress syndrome with anemia at ICU discharge were associated with worsened exercise capacity and more dependency for activities of daily living [25]. Keng et al. reported an increased risk of mechanical ventilator weaning failure among patients with anemia and poor oxygenation at respiratory care center admission [26]. Reade et al. demonstrated that anemia patients (Hgb < 10) are associated with 90-day mortality in cases of hospitalized community-acquired pneumonia [27]. Recently, a large multicenter cohort study reported that COVID-19 patients with anemia are associated with disease severity and mortality [28]. A possible mechanism for the more frequent occurrence of anemia is that inhaled oxygen from the environment crosses the alveolar–capillary membrane into the bloodstream. Most oxygen is bound to Hgb in the red blood cells, although a small amount dissolves in the plasma. Oxygen is then transported from the lungs to the peripheral tissues, where it is removed from the blood and used to promote aerobic cellular metabolism. As the percentage of subjects with Hgb levels of <9 was higher in the HFNC failure group compared with the HFNC success group, the concentration of Hgb in these subjects was likely low, and their ability to carry oxygen became inadequate. Consequently, the hypoxemia of the corresponding subjects did not improve during HFNC therapy, and higher-level support was required.

EIT enables clinicians to measure and observe dynamic regional ventilation distribution [6]. Basile et al. reported reduced GI values in patients receiving HFNC, indicating an improvement in uneven ventilation distribution after HFNC therapy [29]. Pérez-Terán et al. demonstrated that HFNC significantly decreased the respiratory rate and increased end-expiratory lung impedance in patients with respiratory diseases [30]. Recently, Li et al. reported that ventilation distributions among patients with acute respiratory failure during their first hour in the ICU were slightly different but were insignificant in predicting HFNC failure [31]. An effort has been made to use machine learning methods to predict HFNC outcomes using EIT [32]. However, the number of participants was limited to those with effective machine learning. The clinical application of EIT in general respiratory wards has rarely been reported. We found that EIT could be used to monitor patients receiving HFNC therapy in the general ward. Patients with successful HFNC treatment had a more central distribution 2 h after HFNC treatment, lasting for at least 24 h (Figure 2a). Several possible mechanisms may explain this finding. First, HFNC has a low positive end-expiratory pressure effect (e.g., 5 cm H_2O). This may promote ventilation redistribution within a short period in cases of AHRF. Second, HFNC improves patient comfort and provides a stable oxygen concentration under a continuous, high oxygen flow with adequate humidity [4,33]. In the present study, we observed that ventilation in the failure group was distributed slightly towards the ventral region (Figure 2a). Previous studies have suggested that ventilation distribution during spontaneous breathing may be related to diaphragmatic activity [34,35]. The differences found in the CoV might indicate decreased respiratory effort in the failure group, implying possible fatigue of the respiratory muscles. Other EIT-based parameters, such as RVD, did not differ between the subgroups. We suspect that during spontaneous breathing, the inspiratory time is significantly short to obtain stable RVD values. A recent study showed a high coefficient of variation for RVD among healthy subjects [36], which may explain why no significant differences were observed in the present study.

HFNC treatment is generally comfortable for patients in AHRF. In a previous study, clinical staff reported easy use of HFNC devices, whereas patients reported relatively high comfort levels while breathing humidified and preheated air [37]. The benefits of patient tolerance and more reliable FiO_2 delivery due to dead space flushing make HFNC an excellent method for oxygen delivery. Early initiation of HFNC reduces inspiratory effort, thereby reducing pulmonary transvascular pressure and protecting the lungs from patient-inflicted lung damage [38]. A previous study applied the therapeutic benefits of HFNC, namely, the tolerance of long ventilation times, reduced nursing workload, and significant reduction in 90-day mortality previously described in the literature in favor of HFNC, for acute hypoxemic respiration and compared the results with other forms of non-invasive ventilation (e.g., continuous positive airway pressure or bilevel positive airway pressure) [14]. In addition, HFNC therapy improves the respiratory rates, tolerance, and comfort of interstitial lung disease patients with acute hypoxemic respiratory failure who receive supportive care [39].

Our study has some limitations. First, the clinical data were collected from a single medical center. Second, although this was a prospective study, the sample size was relatively small. Third, the causes of the diseases were heterogeneous, which may have caused statistical insignificance in some of the investigated parameters. Further large-scale studies focusing on a single disease for these prognostic factors are necessary to validate these clinical findings.

5. Conclusions

This study suggests that HFNC therapy in general respiratory wards may be a potential rescue therapy for patients with respiratory failure. EIT potentially monitors patients receiving HFNC therapy.

Author Contributions: Conceptualization, C.-H.G. and M.-Y.C.; methodology, Z.Z. and C.-H.G.; software, Z.Z. and T.Z.; validation, C.-H.G., M.-Y.C. and Z.Z.; formal analysis, Z.Z., T.Z. and C.-H.G.; investigation, M.-Y.C. and C.-H.G.; resources, C.-H.G. and M.-Y.C.; data curation, M.-Y.C. and C.-H.G.; writing—original draft preparation, Z.Z., T.Z. and C.-H.G.; writing—review and editing, Z.Z., T.Z. and C.-H.G.; visualization, C.-H.G. and M.-Y.C.; supervision, C.-H.G.; project administration, C.-H.G.;

funding acquisition, C.-H.G. and Z.Z. All authors have read and agreed to the published version of the manuscript.

Funding: This research was funded by Far Eastern Memorial Hospital, Taiwan, grant number FEMH-2020-C-008.

Institutional Review Board Statement: The study was conducted in accordance with the Declaration of Helsinki and approved by the Institutional Review Board of the Far Eastern Memorial Hospital Ethics Committee in Taiwan (FEMH-107139-F).

Informed Consent Statement: Informed consent was obtained from all subjects involved in the study.

Data Availability Statement: All data generated or analyzed in this study are included in the published article. Data are available upon reasonable request to the corresponding author.

Acknowledgments: We acknowledge Ping-Huai Wang from Far Eastern Memorial Hospital (New Taipei City, Taiwan) for recruiting a portion of the patients.

Conflicts of Interest: Z.Z. received a consulting fee from Draeger Medical Company. The other authors declare that they have no competing interests.

References

1. Cirio, S.; Piran, M.; Vitacca, M.; Piaggi, G.; Ceriana, P.; Prazzoli, M.; Paneroni, M.; Carlucci, A. Effects of heated and humidified high flow gases during high-intensity constant-load exercise on severe COPD patients with ventilatory limitation. *Respir. Med.* **2016**, *118*, 128–132. [CrossRef]
2. Spoletini, G.; Alotaibi, M.; Blasi, F.; Hill, N.S. Heated Humidified High-Flow Nasal Oxygen in Adults: Mechanisms of Action and Clinical Implications. *Chest* **2015**, *148*, 253–261. [CrossRef] [PubMed]
3. Zhang, J.; Lin, L.; Pan, K.; Zhou, J.; Huang, X. High-flow nasal cannula therapy for adult patients. *J. Int. Med. Res.* **2016**, *44*, 1200–1211. [CrossRef] [PubMed]
4. Mauri, T.; Alban, L.; Turrini, C.; Cambiaghi, B.; Carlesso, E.; Taccone, P.; Bottino, N.; Lissoni, A.; Spadaro, S.; Volta, C.A.; et al. Optimum support by high-flow nasal cannula in acute hypoxemic respiratory failure: Effects of increasing flow rates. *Intensive Care Med.* **2017**, *43*, 1453–1463. [CrossRef]
5. Sztrymf, B.; Messika, J.; Mayot, T.; Lenglet, H.; Dreyfuss, D.; Ricard, J.D. Impact of high-flow nasal cannula oxygen therapy on intensive care unit patients with acute respiratory failure: A prospective observational study. *J. Crit. Care* **2012**, *27*, 324.e9–324.e13. [CrossRef] [PubMed]
6. Zhao, Z.; Chang, M.Y.; Frerichs, I.; Zhang, J.H.; Chang, H.T.; Gow, C.H.; Moller, K. Regional air trapping in acute exacerbation of obstructive lung diseases measured with electrical impedance tomography: A feasibility study. *Minerva Anestesiol.* **2020**, *86*, 172–180. [CrossRef] [PubMed]
7. Hough, J.L.; Pham, T.M.; Schibler, A. Physiologic effect of high-flow nasal cannula in infants with bronchiolitis. *Pediatr. Crit. Care Med.* **2014**, *15*, e214–e219. [CrossRef]
8. Zhao, Z.; Moller, K.; Steinmann, D.; Frerichs, I.; Guttmann, J. Evaluation of an electrical impedance tomography-based Global Inhomogeneity Index for pulmonary ventilation distribution. *Intensive Care Med.* **2009**, *35*, 1900–1906. [CrossRef]
9. Frerichs, I.; Hahn, G.; Golisch, W.; Kurpitz, M.; Burchardi, H.; Hellige, G. Monitoring perioperative changes in distribution of pulmonary ventilation by functional electrical impedance tomography. *Acta Anaesthesiol. Scand.* **1998**, *42*, 721–726. [CrossRef]
10. Muders, T.; Luepschen, H.; Zinserling, J.; Greschus, S.; Fimmers, R.; Guenther, U.; Buchwald, M.; Grigutsch, D.; Leonhardt, S.; Putensen, C.; et al. Tidal recruitment assessed by electrical impedance tomography and computed tomography in a porcine model of lung injury. *Crit. Care Med.* **2012**, *40*, 903–911. [CrossRef]
11. Saillard, C.; Lambert, J.; Tramier, M.; Chow-Chine, L.; Bisbal, M.; Servan, L.; Gonzalez, F.; de Guibert, J.M.; Faucher, M.; Sannini, A.; et al. High-flow nasal cannula failure in critically ill cancer patients with acute respiratory failure: Moving from avoiding intubation to avoiding delayed intubation. *PLoS ONE* **2022**, *17*, e0270138. [CrossRef]
12. Al-Mazedi, M.S.; Rajan, R.; Al-Jarallah, M.; Dashti, R.; Al Saber, A.; Pan, J.; Zhanna, K.D.; Abdelnaby, H.; Aboelhassan, W.; Almutairi, F.; et al. Neutrophil to lymphocyte ratio and in-hospital mortality among patients with SARS-CoV-2: A retrospective study. *Ann. Med. Surg.* **2022**, *82*, 104748. [CrossRef]
13. Colombo, S.M.; Scaravilli, V.; Cortegiani, A.; Corcione, N.; Guzzardella, A.; Baldini, L.; Cassinotti, E.; Canetta, C.; Carugo, S.; Hu, C.; et al. Use of high flow nasal cannula in patients with acute respiratory failure in general wards under intensivists supervision: A single center observational study. *Respir. Res.* **2022**, *23*, 171. [CrossRef] [PubMed]
14. Frat, J.P.; Thille, A.W.; Mercat, A.; Girault, C.; Ragot, S.; Perbet, S.; Prat, G.; Boulain, T.; Morawiec, E.; Cottereau, A.; et al. High-flow oxygen through nasal cannula in acute hypoxemic respiratory failure. *N. Engl. J. Med.* **2015**, *372*, 2185–2196. [CrossRef] [PubMed]
15. Ma, X.H.; An, M.M.; Yin, F.; Zhang, J.; Peng, M.Y.; Guan, H.; Gong, P. Factors associated with failure of high-flow nasal cannula oxygen therapy in patients with severe COVID-19: A retrospective case series. *J. Int. Med. Res.* **2022**, *50*, 3000605221103525. [CrossRef] [PubMed]

16. Kim, W.Y.; Sung, H.; Hong, S.B.; Lim, C.M.; Koh, Y.; Huh, J.W. Predictors of high flow nasal cannula failure in immunocompromised patients with acute respiratory failure due to non-HIV pneumocystis pneumonia. *J. Thorac. Dis.* **2017**, *9*, 3013–3022. [CrossRef]
17. Demoule, A.; Vieillard Baron, A.; Darmon, M.; Beurton, A.; Geri, G.; Voiriot, G.; Dupont, T.; Zafrani, L.; Girodias, L.; Labbe, V.; et al. High-Flow Nasal Cannula in Critically III Patients with Severe COVID-19. *Am. J. Respir. Crit. Care Med.* **2020**, *202*, 1039–1042. [CrossRef]
18. Kamjai, P.; Hemvimol, S.; Bordeerat, N.K.; Srimanote, P.; Angkasekwinai, P. Evaluation of emerging inflammatory markers for predicting oxygen support requirement in COVID-19 patients. *PLoS ONE* **2022**, *17*, e0278145. [CrossRef]
19. Erdogan, A.; Can, F.E.; Gonullu, H. Evaluation of the prognostic role of NLR, LMR, PLR, and LCR ratio in COVID-19 patients. *J. Med. Virol.* **2021**, *93*, 5555–5559. [CrossRef]
20. Yang, L.; Gao, C.; He, Y.; Wang, X.; Yang, L.; Guo, S.; Chen, J.; He, S.; Sun, Y.; Gao, Y.; et al. The Neutrophil-to-Lymphocyte Ratio is Associated with the Requirement and the Duration of Invasive Mechanical Ventilation in Acute Respiratory Distress Syndrome Patients: A Retrospective Study. *Can. Respir. J.* **2022**, *2022*, 1581038. [CrossRef]
21. Rathod, B.D.; Amle, D.; Khot, R.S.; Prathipati, K.K.; Joshi, P.P. Neutrophil-to-Lymphocyte Ratio as a Predictor of Disease Severity and Mortality in Coronavirus Disease 2019: Prospective Study from Central India. *Cureus* **2022**, *14*, e23696. [CrossRef] [PubMed]
22. Vadi, S.; Pednekar, A.; Suthar, D.; Sanwalka, N.; Ghodke, K.; Rabade, N. Characteristics and Predictive Value of T-lymphocyte Subset Absolute Counts in Patients with COVID-19-associated Acute Respiratory Failure: A Retrospective Study. *Indian. J. Crit. Care Med.* **2022**, *26*, 1198–1203. [CrossRef]
23. Hayden, S.J.; Albert, T.J.; Watkins, T.R.; Swenson, E.R. Anemia in critical illness: Insights into etiology, consequences, and management. *Am. J. Respir. Crit. Care Med.* **2012**, *185*, 1049–1057. [CrossRef] [PubMed]
24. Khamiees, M.; Raju, P.; DeGirolamo, A.; Amoateng-Adjepong, Y.; Manthous, C.A. Predictors of extubation outcome in patients who have successfully completed a spontaneous breathing trial. *Chest* **2001**, *120*, 1262–1270. [CrossRef] [PubMed]
25. Warner, M.A.; Hanson, A.C.; Frank, R.D.; Schulte, P.J.; Go, R.S.; Storlie, C.B.; Kor, D.J. Prevalence of and Recovery from Anemia Following Hospitalization for Critical Illness among Adults. *JAMA Netw. Open* **2020**, *3*, e2017843. [CrossRef]
26. Keng, L.T.; Chung, K.P.; Lin, S.Y.; Liang, S.K.; Cheng, J.C.; Chen, I.C.; Chen, Y.F.; Chang, H.T.; Hsu, C.L.; Jerng, J.S.; et al. Significant Clinical Factors Associated with Long-term Mortality in Critical Cancer Patients Requiring Prolonged Mechanical Ventilation. *Sci. Rep.* **2017**, *7*, 2148. [CrossRef]
27. Reade, M.C.; Weissfeld, L.; Angus, D.C.; Kellum, J.A.; Milbrandt, E.B. The prevalence of anemia and its association with 90-day mortality in hospitalized community-acquired pneumonia. *BMC Pulm. Med.* **2010**, *10*, 15. [CrossRef]
28. Veronese, N.; Segala, F.V.; Carruba, L.; La Carrubba, A.; Pollicino, F.; Di Franco, G.; Guido, G.; Cormio, M.; Lugli, A.; De Santis, L.; et al. Anemia as a risk factor for disease progression in patients admitted for COVID-19: Data from a large, multicenter cohort study. *Sci. Rep.* **2023**, *13*, 9035. [CrossRef]
29. Basile, M.C.; Mauri, T.; Spinelli, E.; Dalla Corte, F.; Montanari, G.; Marongiu, I.; Spadaro, S.; Galazzi, A.; Grasselli, G.; Pesenti, A. Nasal high flow higher than 60 L/min in patients with acute hypoxemic respiratory failure: A physiological study. *Crit. Care* **2020**, *24*, 654. [CrossRef]
30. Perez-Teran, P.; Marin-Corral, J.; Dot, I.; Sans, S.; Munoz-Bermudez, R.; Bosch, R.; Vila, C.; Masclans, J.R. Aeration changes induced by high flow nasal cannula are more homogeneous than those generated by non-invasive ventilation in healthy subjects. *J. Crit. Care* **2019**, *53*, 186–192. [CrossRef]
31. Li, Z.; Zhang, Z.; Xia, Q.; Xu, D.; Qin, S.; Dai, M.; Fu, F.; Gao, Y.; Zhao, Z. First Attempt at Using Electrical Impedance Tomography to Predict High Flow Nasal Cannula Therapy Outcomes at an Early Phase. *Front. Med.* **2021**, *8*, 737810. [CrossRef] [PubMed]
32. Yang, L.; Li, Z.; Dai, M.; Fu, F.; Moller, K.; Gao, Y.; Zhao, Z. Optimal machine learning methods for prediction of high-flow nasal cannula outcomes using image features from electrical impedance tomography. *Comput. Methods Programs Biomed.* **2023**, *238*, 107613. [CrossRef] [PubMed]
33. Mauri, T.; Turrini, C.; Eronia, N.; Grasselli, G.; Volta, C.A.; Bellani, G.; Pesenti, A. Physiologic Effects of High-Flow Nasal Cannula in Acute Hypoxemic Respiratory Failure. *Am. J. Respir. Crit. Care Med.* **2017**, *195*, 1207–1215. [CrossRef] [PubMed]
34. Ischaki, E.; Pantazopoulos, I.; Zakynthinos, S. Nasal high flow therapy: A novel treatment rather than a more expensive oxygen device. *Eur. Respir. Rev.* **2017**, *26*, 170028. [CrossRef]
35. Zhao, Z.; Peng, S.Y.; Chang, M.Y.; Hsu, Y.L.; Frerichs, I.; Chang, H.T.; Moller, K. Spontaneous breathing trials after prolonged mechanical ventilation monitored by electrical impedance tomography: An observational study. *Acta Anaesthesiol. Scand.* **2017**, *61*, 1166–1175. [CrossRef]
36. Yang, L.; Dai, M.; Cao, X.; Moller, K.; Dargvainis, M.; Frerichs, I.; Becher, T.; Fu, F.; Zhao, Z. Regional ventilation distribution in healthy lungs: Can reference values be established for electrical impedance tomography parameters? *Ann. Transl. Med.* **2021**, *9*, 789. [CrossRef]
37. Spoletini, G.; Hill, N.S. High-flow nasal oxygen versus noninvasive ventilation for hypoxemic respiratory failure: Do we know enough? *Ann. Thorac. Med.* **2016**, *11*, 163–166. [CrossRef]

38. Marini, J.J.; Gattinoni, L. Management of COVID-19 Respiratory Distress. *JAMA* **2020**, *323*, 2329–2330. [CrossRef]
39. Koyauchi, T.; Hasegawa, H.; Kanata, K.; Kakutani, T.; Amano, Y.; Ozawa, Y.; Matsui, T.; Yokomura, K.; Suda, T. Efficacy and Tolerability of High-Flow Nasal Cannula Oxygen Therapy for Hypoxemic Respiratory Failure in Patients with Interstitial Lung Disease with Do-Not-Intubate Orders: A Retrospective Single-Center Study. *Respiration* **2018**, *96*, 323–329. [CrossRef]

biomedicines

MDPI

Article

Long-Term Radiological Pulmonary Changes in Mechanically Ventilated Patients with Respiratory Failure due to SARS-CoV-2 Infection

Mircea Stoian [1], Adina Roman [2,*], Alina Boeriu [2], Danusia Onișor [2], Sergio Rareș Bandila [3], Dragoș Florin Babă [4], Iuliu Cocuz [5], Raluca Niculescu [5], Anamaria Costan [6], Sergiu Ștefan Laszlo [7,*], Dragoș Corău [7] and Adina Stoian [5]

1 Department of Anesthesiology and Intensive Care, George Emil Palade University of Medicine, Pharmacy, Sciences and Technology of Târgu Mureș, 540139 Targu Mures, Romania; mircea.stoian@umfst.ro
2 Gastroenterology Department, George Emil Palade University of Medicine, Pharmacy, Sciences and Technology of Târgu Mureș, 540142 Targu Mures, Romania; aboeriu@gmail.com (A.B.); danusia.onisor@umfst.ro (D.O.)
3 Orthopedic Surgery and Traumatology Service, Marina Baixa Hospital, Av. Alcade En Jaume Botella Mayor, 03570 Villajoyosa, Spain; sergiob1976@gmail.com
4 Department of Cell and Molecular Biology, George Emil Palade University of Medicine, Pharmacy, Sciences and Technology of Târgu Mureș, 540142 Targu Mures, Romania; dragos-florin.baba@umfst.ro
5 Department of Pathophysiology, George Emil Palade University of Medicine, Pharmacy, Sciences and Technology of Târgu Mureș, 540136 Targu Mures, Romania; iuliu.cocuz@umfst.ro (I.C.); raluca.niculescu@umfst.ro (R.N.); adina.stoian@umfst.ro (A.S.)
6 Faculty of Medicine, George Emil Palade University of Medicine, Pharmacy, Sciences and Technology of Târgu Mureș, 540142 Targu Mures, Romania; anamariacostan97@yahoo.com
7 Intensive Care Unit, Mureș County Hospital, Street Gheorghe Marinescu no 1, 540136 Targu Mures, Romania; dragos.corau@gmail.com
* Correspondence: adina.roman91@gmail.com (A.R.); laszlo.sergiu.stefan@gmail.com (S.Ș.L.)

Abstract: From the first reports of SARS-CoV-2, at the end of 2019 to the present, the global mortality associated with COVID-19 has reached 6,952,522 deaths as reported by the World Health Organization (WHO). Early intubation and mechanical ventilation can increase the survival rate of critically ill patients. This prospective study was carried out on 885 patients in the ICU of Mureș County Clinical Hospital, Romania. After applying inclusion and exclusion criteria, a total of 54 patients were included. Patients were monitored during hospitalization and at 6-month follow-up. We analyzed the relationship between invasive mechanical ventilation (IMV) and non-invasive mechanical ventilation (NIMV) and radiological changes on thoracic CT scans performed at 6-month follow-up and found no significant association. Regarding paraclinical analysis, there was a statistically significant association between patients grouped by IMV and ferritin level on day 1 of admission ($p = 0.034$), and between patients grouped by PaO_2/FiO_2 ratio with metabolic syndrome ($p = 0.03$) and the level of procalcitonin ($p = 0.01$). A significant proportion of patients with COVID-19 admitted to the ICU developed pulmonary fibrosis as observed at a 6-month evaluation. Patients with oxygen supplementation or mechanical ventilation require dynamic monitoring and radiological investigations, as there is a possibility of long-term pulmonary fibrosis that requires pharmacological interventions and finding new therapeutic alternatives.

Keywords: severe acute respiratory syndrome coronavirus 2 (SARS-CoV-2); corona virus disease 2019 (COVID-19); acute respiratory distress syndrome (ARDS); intensive care unit (ICU); invasive ventilation; pulmonary fibrosis; respiratory failure

1. Introduction

From the first reports of a novel coronavirus (severe acute respiratory syndrome coronavirus 2 [SARS-CoV-2]) at the end of 2019 to the present, 768,560,727 confirmed cases of infection have been recorded, and the global mortality associated with corona virus disease 2019 (COVID-19) has reached 6,952,522 deaths as reported by the World Health Organization (WHO) [1]. Even though most cases with mild symptoms recover in 1–2 weeks, some cases evolve unpredictably, with manifestations ranging from a lack of symptoms to complications in different organs; some cases even evolve towards severe respiratory failure requiring mechanical ventilation, frequently resulting in the death of the patient [2–6]. Like other coronaviruses that have caused severe outbreaks, SARS-CoV-2 primarily affects the respiratory system. COVID-19 was recognized early on based on its pulmonary manifestations, but the medical world quickly realized that it was a systemic disease that affected multiple organs, such as the nervous, cardiovascular, hematopoietic, gastrointestinal, and renal systems [7–12].

Histopathological and imaging abnormalities characteristic of lung fibrosis were found in survivors of the 2003 severe acute respiratory syndrome coronavirus 1 (SARS-CoV-1) outbreak [13]. Recent studies published after the SARS-CoV-2 pandemic also attest to the development of pulmonary fibrosis in patients who presented with lung damage or acute respiratory distress syndrome (ARDS) [11,14], suggesting that SARS viruses have greater potential to induce a lung fibrotic proliferative response than other respiratory viruses [15]. Sturgill et al. published a study in 2023 in which they found that the occurrence of fibrotic changes on chest CT imaging for COVID-19 survivors of pneumonia with ARDS is higher compared to patients with non-COVID ARDS etiologies. This fact is assumed to result from the pathophysiological aspects of the evolution of SARS-CoV2 infection to which the pharmaceutical treatment and the particularities of ventilation in these patients are added [16]. The most characteristic radiographic changes in pulmonary fibrosis include traction bronchiectasis, reticulations, honeycombing, and ground-glass opacities [15].

The alveoli–capillary membrane consists of the following structures:

(1) The alveolar epithelium situated on its basement membrane.
(2) The capillary endothelium that lays on its basement membrane.
(3) A thin interstitium separating the two basement membranes, a connective tissue containing fibroblasts, macrophages, collagen, and elastic fibers [17].

Type I alveolar epithelial cells (AECIs) surround the alveolar air space and participate in gas exchange, while type II alveolar epithelial cells (AECIIs) are responsible for surfactant production and can differentiate into AECIs to repair the defects [18].

ARDS is characterized by bilateral pulmonary infiltrates with an acute onset and severe evolution, characterized by hypoxemia and non-cardiac pulmonary edema that require mechanical ventilation (sometimes prolonged) as supportive therapy [19,20]. The main stages of ARDS evolution are the following:

(1) An exudative, early phase with disruption of the alveolocapillary membrane and edematous flooding of the alveolar spaces, followed by
(2) A proliferative phase, with clearance of exudative fluid and an attempt to reestablish the alveolar barrier; then, in patients who survive,
(3) A fibrotic phase characterized by fibroproliferation, with some patients undergoing excessive fibrotic process [20,21].

Early intubation and mechanical ventilation can increase the survival rate of critically ill patients with COVID-19 ARDS [22,23]. However, mechanical ventilation in the intensive care unit (ICU), a significant number of patients admitted in the ICU, and crowding in these units are associated with a higher mortality risk [2]. Different outcomes of mechanical ventilation in COVID-19 patients have been reported in publications worldwide [24]. In a study conducted in China, out of 344 patients hospitalized in the ICU, 100 (29.1%) required invasive mechanical ventilation [25]. Barotrauma associated with mechanical ventilation for respiratory failure due to COVID-19 was found in 40% of cases. These complications can occur even when protective mechanical ventilation recommendations are followed [26].

Data regarding long-term respiratory changes in mechanically ventilated patients with invasive or non-invasive positive pressure are limited [27]. Our study aimed to analyze and describe the main risk factors associated with the development of post-COVID-19 pulmonary fibrosis and gain a better understanding of its pathophysiology. A secondary objective was to analyze the relationship between mechanical ventilation and the degree of long-term pulmonary lesions in patients with a medium-to-severe form of COVID-19 treated in the ICU of Mureș Clinical County Hospital.

Is this subject still relevant? Unfortunately, the risk of new outbreaks of COVID-19 remains, even though the pandemic has been officially declared to have ended. Thus, we consider the publication of data about the incidence of pulmonary fibrosis in COVID-19 survivors who suffered from ARDS and the effect of mechanical ventilation on long-term respiratory function to be important, given the small number of articles published in the literature regarding this topic thus far.

2. Materials and Methods

The study protocol was approved by the Ethics Committee of the Mureș Clinical County Hospital. All procedures were conducted according to the principles of the Helsinki Declaration.

This prospective study was carried out in the ICU of Mureș County Clinical Hospital, Tîrgu Mureș, Romania, which contains 20 beds for critically ill patients. This ward functioned as an ICU for COVID-19 patients during the pandemic period. The study, lasting 1 year and 9 months, was conducted between April 2020 and December 2021. A total of 885 SARS-CoV-2-positive patients were admitted to the ICU. After reviewing the inclusion and exclusion criteria and eliminating the deceased patients, 91 were included in the study. Patients were monitored from the day of hospital discharge until 6 months after discharge. Six months after ICU discharge, the participants underwent a non-contrast thoracic CT scan. From this group, 35 patients were excluded because they did not attend the periodic evaluations and provide all the requested data or died.

Thus, the final study group included 56 adult patients (Figure 1) diagnosed with a medium/severe form of respiratory failure that fulfilled the criteria for ARDS according to the Berlin definition [19] and was caused by SARS-CoV-2 infection, with a positive reverse transcription polymerase-chain-reaction (RT-PCR) swab test, who were mechanically ventilated for more than 48 h during the hospitalization period (33 with non-invasive mechanical ventilation [NIMV] and 23 with invasive mechanical ventilation [IMV]). Pulmonary fibrosis was diagnosed according to radiological criteria by performing a computerized tomography (CT) scan 6 months after discharge and categorized according to severity.

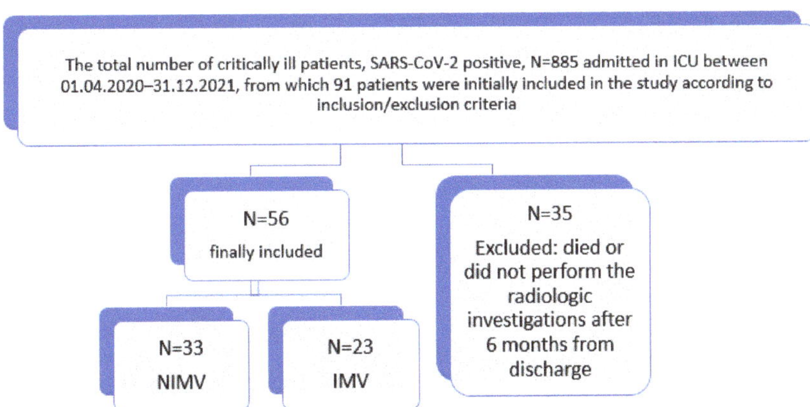

Figure 1. Flow diagram of study participants.

Inclusion criteria: Adult patients over 18 years of age who gave their consent to enter the study, with a positive RT-PCR SARS-CoV-2 diagnostic test, moderate or severe respiratory failure due to SARS-CoV-2 infection, and invasive or non-invasive mechanical ventilation for more than 48 h during hospitalization and who underwent a radiological investigation at 6 months after discharge.

Exclusion criteria: Patients under 18 years of age, those who did not provide informed consent to enter the study, those who did not agree to the publication of their data, those who died during the first 6 months after discharge, those who did not attend periodic evaluations, and those with incomplete data.

2.1. Data Collection

Patient data collected during hospitalization included sex, age, urban or rural origin, existing comorbidities, Quick Sequential Organ Failure Assessment (qSOFA) score at admission [28], acute physiology and chronic health evaluation II (APACHE II) score completed in the first 24 h after admission in the ICU [29], the ratio of arterial oxygen partial pressure (PaO_2) to fractional inspired oxygen (FiO_2) (PaO_2/FiO_2), IMV/NIMV, radiological investigations (thoracic CT scan), and laboratory parameters (white blood cell [WBC] count, serum ferritin level, D-dimer level, fibrinogen, C-reactive protein [CRP], procalcitonin, lactate dehydrogenase [LDH], and serum creatinine level). Those who received interleukin (IL) 6 (IL-6) blockers were also registered.

Other collected data included the number of days spent in the ICU, the type of positive-pressure ventilation applied, and the number of hours spent on the ventilator during hospitalization.

The patients included in the study were evaluated 6 months after discharge, and the examination performed on this occasion comprised a radiological investigation (thoracic CT scan). Depending on the severity of the radiological changes, the severity of post-discharge pulmonary involvement was classified as follows:

(1) Mild-to-moderate forms in those who presented pulmonary radiological changes of 10–50%.
(2) Severe forms in patients presenting more than 50% pulmonary radiological changes.

After performing data collection, the patients who were administered mechanical ventilation were classified into two groups:

(1) Those with NIMV
(2) Those with IMV

2.2. Statistical Analysis

Software applications such as IBM SPSS Statistics v26 (New York, NY, USA) and Microsoft Office Excel 2019 (Washington, DC, USA) were used for statistical analysis. The statistical analyses consisted of an assessment of parametric variables (ANOVA test), describing the data as continuous (mean, standard deviation [SD], median, min/max) depending on their distribution. Quantitative variables were correlated using the Pearson correlation coefficient (rho), with alpha set at 0.05. A p-value $\leq$ 0.05 was considered significant. To assess the correlation between the distributions of the categorical variables, contingency tables, and the Chi-squared test were used.

3. Results

3.1. Participant Characteristics

The study was conducted on 56 adult patients (over 18 years of age) diagnosed with a medium-to-severe form of ARDS according to the Berlin definition caused by SARS-CoV-2, with a positive RT-PCR test, who were mechanically ventilated for more than 48 h. Thirty-three (58.9%) patients received NIMV and 23 (41.07%) received IMV.

The mean age of the included patients was 56.54 ± 12.83 years (a range of 24–81 years). The mean age of the patients with IMV was 58.70 ± 12.11 years, and those with NIMV were 55.03 ± 13.28 years, without significant differences between these two groups (p = 0.297). The gender distribution was 33 men (58.9%) and 23 women (41.1%). We did not find a significant association between the presence of pulmonary fibrosis and the patients' gender.

The age, sex, and environment (urban/rural) of the participants were not statistically significantly correlated with the APACHE II or qSOFA severity scores or the presence or absence of pulmonary fibrosis.

Regarding rural or urban origin, 16 patients (28.6%) came from rural areas, and 40 patients (71.4%) from urban areas. We did not find a significant association between the presence of pulmonary fibrosis and the patients' origin (rural or urban) (p = 1.00).

3.2. Associated Pre-Existing Comorbidities

Out of the total 56 patients, 49 (87.50%) had associated conditions such as pre-existing pulmonary diseases (13; 23.21%), cardiovascular diseases (26; 46.43%), neurological diseases (4; 7.14%), kidney diseases (8; 14.29%), hematological diseases (12; 21.43%), rheumatological diseases (8; 14.29%), metabolic syndrome (31; 55.36%), and endocrinological diseases (5; 8.93%). Upon analyzing the correlations between the number of days spent in the ICU and the presence of comorbidities, we obtained the following results: an increased number of hospitalization days correlated with the presence of kidney diseases (p = 0.04) but not with the presence of neurological diseases (p = 1.00), pulmonary diseases (p = 0.10), hematological diseases (p = 0.68), rheumatological diseases (p = 0.64), metabolic syndrome (p = 0.51) or endocrinological diseases (p = 0.57).

The presence of kidney disease was correlated with the presence of mild/moderate pulmonary fibrosis (p = 0.04).

The patients' characteristics and associated comorbidities are shown in Tables 1 and 2.

Table 1. Patients' age and severity score.

Variable (n = 56)	Mean ± Standard Deviation (Min–Max)
Age (years)	56.54 ± 12.83 (24–81)
APACHE II	22.41 ± 9.48 (10–45)
q-SOFA	1.96 ± 0.73 (1–3)

Table 2. Patients' clinical characteristics and associated comorbidities.

Variable (n = 56)	Count (Number of Patients)	Percentage (%)
Sex (Male:Female)	33:23	58.93%:41.07%
ICU admission days		
<10 days	45	80.36%
≥10 days	11	19.64%
Patients with mechanical ventilation		
<100 h	15	26.79%
≥100 h	41	73.21%
Patients with NIMV vs. IMV		
NIMV:IMV	33:23	58.93%:41.07%
Oxygen dependency at discharge	28	50%
Pulmonary fibrosis		
No fibrosis	7	12.50%
Minor–Moderate	38	67.86%
Severe	11	19.64%
PaO_2/FiO_2		
<100	15	26.79%
100–200	32	57.14%
200–300	8	14.29%
Comorbidities		
Cardiovascular	26	46.43%
Pulmonary	13	23.21%
Neurological	4	7.14%
Renal	8	14.29%
Hematological	12	21.43%

Table 2. *Cont.*

Variable (n = 56)	Count (Number of Patients)	Percentage (%)
Rheumatological	8	14.29%
Metabolic syndrome	31	55.36%
Endocrine	5	8.93%
Other comorbidities	26	46.43%

3.3. The Duration of Mechanical Ventilation

Fifteen patients (26.79%) required less than 100 h of mechanical ventilation (IMV/NIMV), and 41 patients (73.21%) required more than 100 h of mechanical ventilation (IMV/NIMV). A total of 45 patients (80.36%) were hospitalized in the ICU for less than 10 days, and 11 (19.64%) patients were administered ventilation for more than 10 days.

The association between patients grouped by the mean days of hospitalization and the presence of pulmonary fibrosis was not statistically significant. IMV or NIMV were not correlated with the APACHE or SOPHA severity scores.

The associations between patients grouped by mean days of hospitalization and the number of hours of mechanical ventilation (p = 0.02, ANOVA) and the type of ventilation—NIMV (p = 0.01, ANOVA) and IMV (p = 0.01, ANOVA), were statistically significant.

Upon evaluating the association between the number of hours of mechanical ventilation and the occurrence of radiological changes suggestive of pulmonary fibrosis on the thoracic CT scan performed 6 months after discharge, we did not observe a statistically significant association between the number of hours of mechanical ventilation and the presence of mild pulmonary fibrosis (p = 1.00, ANOVA) or moderate/severe pulmonary fibrosis (p = 0.14, ANOVA).

3.4. Number of Days Spent in the ICU

A total of 45 patients (80.36%) were admitted to the ICU for less than 10 days, and 11 patients (19.64%) required a more extended hospitalization in the ICU (more than 10 days). There was no statistically significant association between patients grouped by the mean duration of hospitalization in the ICU and the presence of pulmonary fibrosis at 6 months, as evidenced by radiographic tests (p = 0.32, Chi-squared test). Although the degree of fibrosis was not significantly correlated, there was no association between patients grouped by the mean days of hospitalization and the level of SpO_2 (p = 0.02, ANOVA test) and PaO_2/FiO_2 ratio of 100–200 (p = 0.01, ANOVA test).

Upon assessing the relationship between the patients grouped by the mean duration of hospitalization in the ICU and the laboratory values of some biochemical parameters (CRP, ferritin, fibrinogen, procalcitonin, D-dimers, LDH, glucose, aspartate aminotransferase [AST], alanine transaminase [ALT], blood urea nitrogen, and creatinine) determined at admission, there were no statistically significant correlations found.

3.5. PaO_2/FiO_2 Ratio

Regarding the PaO_2/FiO_2 ratio at the time of ICU admission, we found the following distribution of patients: 15 patients (26.79%) with $PaO_2/FiO_2 < 100$, 32 patients (57.14%) with PaO_2/FiO_2 between 100 and 200, eight patients (14.29%) with PaO_2/FiO_2 between 200 and 300, and no patients (0%) with PaO_2/FiO_2 >300.

Evaluating the PaO_2/FiO_2 ratio at admission to the ICU and the hours of mechanical ventilation showed no association between the two. Patients with a PaO_2/FiO_2 ratio at admission of below 100 mm Hg showed alterations in the radiological investigations performed 6 months after ICU discharge, but the association was not statistically significant.

The associations between the subgroup of patients grouped by PaO_2/FiO_2 ratio and the administration of IL-6 blocker treatment (p = 0.04, Chi-squared test) metabolic syndrome (p = 0.03, Chi-squared test) were statistically significant. Metabolic syndrome was defined as abdominal obesity, dyslipidemia, high blood pressure, and elevated fasting glucose.

Regarding paraclinical examination, the associations between the subgroup of patients grouped by PaO_2/FiO_2 ratio and the serum level of urea at admission (p = 0.007, Chi-squared test), the level of procalcitonin (p = 0.01, Chi-squared test), and the levels of AST and ALT at admission (p = 0.002 and 0.02, respectively, Chi-squared test), were statistically significant.

The association between the subgroup of patients grouped by PaO_2/FiO_2 ratio and the presence of pulmonary fibrosis, regardless of severity, was not statistically significant.

3.6. Radiological Evaluation of Patients 6 Months after Discharge

All the patients underwent a thoracic CT scan at 6 months after discharge. Of these patients, seven (12.50%) presented with no fibrosis, 38 (67.86%) presented with mild/moderate fibrotic changes, and 11 (19.64%) presented with severe pulmonary fibrosis. Furthermore, 87.5% of the patients treated with IMV or NIMV in the ICU presented fibrotic changes in the radiological investigations performed 6 months after ICU discharge.

All CT scans performed at admission and 6-month follow-up were performed on CT Siemens Somatom go. Top 128 slice provides detailed imaging of the thoracic examinations performed in Mures County Clinical Hospital.

The top image describes in a native CT scan, coronal section, pulmonary window—the presence of ground-glass opacity (as pointed by arrows in Figure 2—top), parenchymal and subpleural bands, and reticular abnormality (Figure 2—top).

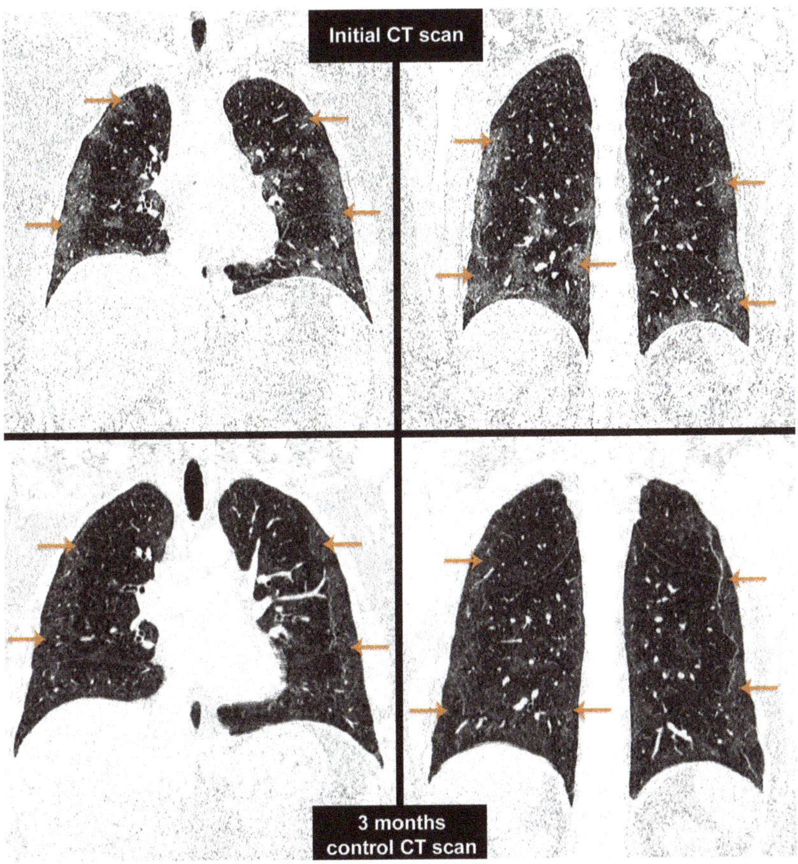

Figure 2. CT scan—initial vs. control (3-month follow-up after hospitalization).

CT at 3-month follow-up after hospitalization, native CT scan, coronal section, pulmonary window—shows a patchy nodular consolidation (as pointed by arrows in Figure 2—bottom), moderate residual ground-glass opacity, and a honeycombing sign with architectural distortion (Figure 2—bottom).

3.7. Evaluation of the Patients with NIMV

We analyzed the relationship between NIMV and radiological changes on the thoracic CT scan performed 6 months after discharge from the ICU and found no significant association. The association between the subgroup of patients grouped by NIMV and the ferritin level at admission was statistically significant ($p = 0.03$, Chi-squared test), but there were no statistically significant correlations between the degree of fibrosis and other laboratory tests. There were no statistically significant differences between NIMV and IMV regarding pulmonary fibrosis or the number of hours of mechanical ventilation.

3.8. Evaluation of the Patients with IMV

We analyzed the relationship between IMV and radiological changes on the thoracic CT scan performed 6 months after discharge from the ICU and found no significant association. IMV was also not correlated with pulmonary fibrosis or the status of O_2-dependent participants at discharge. The association between patients grouped by IMV and the ferritin level on day 1 of admission ($p = 0.034$, Chi-squared test) was statistically significant, but not with ferritin on day 3 of admission ($p = 0.44$, Chi-squared test), as seen in Table 3.

Table 3. Paraclinical examinations performed on day 1 and day 3 of admission.

	Cut-Off Value	Day 1 (Number of Patients)	Day 3 (Number of Patients)	Evolution
CRP (mg/dL)	<0.5	7	7	↔
	≥0.5	49	49	↔
Fibrinogen (mg/dL)	<196	3	3	↔
	196–372	14	16	↗
	>372	39	37	↘
Ferritin (ng/mL)	<300	7	8	↗
	≥300	49	48	↘
Procalcitonin (ng/mL)	<0.5	31	32	↗
	≥0.5	25	24	↘
D-Dimers	Negative	27	23	↘
	Positive	29	33	↗
LDH (U/L)	<220	9	12	↗
	≥220	47	44	↘
Leucocytes (/mm^3)	<4000	3	2	↘
	4000–10000	11	17	↗
	>10000	42	37	↘
Lymphocytes (%)	<25%	47	46	↘
	25–45%	4	9	↗
	>45%	5	1	↗
Neutrophils (%)	<40%	3	1	↗
	40–75%	15	10	↘
	>75%	38	45	↗
AST (U/L)	<34	19	20	↗
	≥34	37	36	↘
ALT (U/L)	<55	27	27	↔
	≥55	29	29	↔

Table 3. *Cont.*

	Cut-Off Value	Day 1 (Number of Patients)	Day 3 (Number of Patients)	Evolution
Creatinine (mg/dL)	<1	38	45	↗
	≥1	18	11	↘
Urea (mg/dL)	<55	29	23	↘
	≥55	27	33	↗
Glucose (mg/dL)	<80	0	1	↗
	80–110	15	11	↘
	>110	41	44	↗

↔ = stationary; ↗ = increased; ↘ = decreased.

3.9. Paraclinical Laboratory Examinations

Regarding the studied paraclinical parameters, we established a statistically significant association between patients grouped by CRP level and LDH at admission ($p = 0.03$, Chi-squared test), using cut-off values as seen in Table 3.

The association between patients grouped by ferritin levels on day 1 of admission and ferritin levels on day 3 ($p = 0.005$, Chi-squared test) was statistically significant.

The association between patients grouped by procalcitonin levels and creatinine levels at admission ($p = 0.04$, Chi-squared test) was statistically significant.

Regarding liver panel tests, we established a statistically significant association between ALT levels on both day 1 and day 3 of admission and AST levels on day 1 and day 3 of admission ($p = 0.01$, Chi-squared test).

There was no statistically significant association between patients grouped by leucocyte count and the presence of pulmonary fibrosis, the number of hours of mechanical ventilation, or the type of ventilation used.

There was no statistically significant association between the patients grouped by the status of O_2-dependent participants at discharge and the degree of pulmonary fibrosis (mild/moderate or severe). The evolution of different paraclinical examinations performed on the patients is presented in Table 3.

4. Discussion

4.1. Risk Factors for Pulmonary Fibrosis

The lungs of COVID-19 patients are damaged due to the viral infection that evolves severely with cytokine storm and the overlapping of mechanical ventilation and bacterial coinfections. Several other factors contribute to pulmonary fibrosis in patients admitted to the ICU, including post-thrombotic events with microvascular damage and alveoli–capillary membrane dysfunction, post-ischemic effects, corticosteroid use, and prolonged ICU stay [11,14]. All these factors act simultaneously, determining the progression of the injury towards fibroproliferation.

4.1.1. ARDS

ARDS was present in all patients admitted to ICU (100%). ARDS is one of the most important predictors of mortality in COVID-19 patients [21]. AECIIs express more abundant angiotensin-converting enzyme 2 (ACE2) receptors, resulting in the higher infectivity of these cells, which may explain the defective surfactant production and re-epithelization of the broncho-alveolar epithelium after injury [30–33]. Epithelial injury is accompanied by endothelial injury, microthrombosis, subsequent pulmonary edema, and vascular leak [21]. ARDS is reported to occur in up to 14% of COVID-19 cases and is a potentially fatal condition [34].

A hallmark of COVID-19 is extensive alveolar epithelial cell injury with secondary fibrotic proliferation, indicating the potential for chronic alveolar and vascular remodeling leading to pulmonary fibrosis and /or pulmonary hypertension [35–37]. According to the literature, pulmonary fibrosis can develop immediately after discharge or in the subsequent weeks, but in some cases, it improves in the months following COVID-19 recovery [11].

Studies estimate that 70–80% of the survivors of severe COVID-19 continue to suffer from long-term postinfectious complications [3,38], of which pulmonary fibrosis is among the most severe and frequently reported [11,39]. The exact prevalence of this sequela is not yet fully established, but some studies reported a prevalence of 62% after SARS-CoV-2 infection [40].

4.1.2. Hypercoagulability

D-dimers were increased in 29 out of 56 patients on the first day of ICU admission, and the number increased to 33 out of 56 on the third day. COVID-19 induces a prothrombotic state, leading to micro- and macrovascular thrombosis, including pulmonary artery thrombi and fibrin microthrombi in small pulmonary arteries [41]. Epidemiological studies have shown a link between thrombotic vascular events and idiopathic pulmonary fibrosis (IPF), with a central role played by the coagulation and fibrinolysis systems in the wound healing and repair processes. A correlation was observed between IPF and the prothrombotic state, but trials with anticoagulation treatment regarding both the progression of fibrotic and thrombotic risks are inconsistent [42]. The logical question that arises in the context of coagulopathies and the hypercoagulable status with inflammatory thrombosis in the pulmonary vasculature associated with COVID-19 is whether this also contributes to the occurrence of pulmonary fibrosis.

4.1.3. Role of Oxygen

All our patients required oxygen supplementation during hospitalization. Moreover, 50% of them presented with oxygen dependency at discharge. The impairment of gas exchange in ARDS culminates in significant hypoxemia. Hypoxemia secondary to lung disease activates hypoxia-inducible factor 1-alpha (HIF-1-α), stimulating tissue fibrosis and activating fibroblasts and ECM deposition [41]. In IPF, the dysregulated expression of HIF-1-α augments myofibroblast differentiation [43]. On the other hand, it was proven that supplemental oxygen used to treat newborns with respiratory disorders and exposure to high-concentration oxygen are associated with inflammation, acute lung injury, and lung fibrosis [44]. The questions that arise in COVID-19 are "How much oxygen is required?" and "How best to titrate it to be useful?" Prolonged oxygen therapy can also increase hyperoxia-induced ROS levels, leading to protein denaturation and the breaking down of nucleic acids [45]. Typically, in viral respiratory diseases, ROS enhance the phagocytosis and activity of immune cells, but excessive oxygenation can cause an imbalance, with extra amounts of ROS that alter redox homeostasis and contribute to alveolocapillary membrane destruction [46]. This combination of SARS-CoV-2 infection and excess ROS production exacerbates cell apoptosis of the alveolar epithelium [47]. An overcorrection of hypoxemia can produce oxygen-induced ARDS and VILI [48].

4.1.4. Age and Sex

Some studies associate the occurrence of pulmonary fibrosis with advanced age, an underlying history of diabetes mellitus, and respiratory and cardiovascular diseases [14,49]. In our study, the mean age of fibrotic patients was 57.08 $\pm$ 12.63 years, and that of non-fibrotic patients was 52.71 $\pm$ 14.60 years. These values are similar to the ages reported by Amin et al. (59 years in those with fibrotic changes compared with 48.5 years in those without fibrotic changes) [11]. Increased age is associated with lung parenchymal stiffening and the profibrotic potential of lung fibroblasts and is considered a major risk factor for the occurrence and progression of pulmonary fibrosis [14,50]. Experimental studies using mouse models reported that fibroblasts and myofibroblasts were more resistant to apoptosis in older animals [51]. Cilli et al. showed that older people with pre-existing chronic respiratory diseases such as IPF represent a vulnerable population group for COVID-19 with a more severe course of the disease and increased mortality rates [52].

The data regarding the association between sex and pulmonary fibrosis after COVID-19 are contradictory, with some studies suggesting a higher prevalence of pulmonary fibrosis in men, while others do not highlight differences between the sexes in favor of men [11,53]. Other studies also correlate the occurrence of pulmonary fibrosis with risk factors such as older age, mechanical ventilation, and the female sex instead of the male sex [54,55]. In our study, no higher prevalence of pulmonary fibrosis in either female or male patients was observed; thus, we could not establish sex or even age as a risk factor for developing pulmonary fibrosis. The mean age of patients included in the study who developed pulmonary fibrosis is similar to that reported in other studies where age proved to be a risk factor for the occurrence of pulmonary fibrosis, and the percentage of men (58.9%) included in the study was higher than that of women.

4.1.5. Associated Comorbidities

Amin et al. reported that of all the comorbidities tested, only chronic obstructive pulmonary disease (COPD) was linked to an increased risk of pulmonary fibrosis [11]. Other studies reported associations between hypertension and pulmonary fibrosis [56].

In our study, mild/moderate pulmonary fibrosis was correlated with renal disease, whereas cardiovascular and pulmonary comorbidities were not correlated with the presence of pulmonary fibrosis. Patients with renal disease also presented with a greater length of hospitalization. We found a correlation between chronic kidney disease and the presence of mild/moderate pulmonary fibrosis, and the studies also suggest that it is correlated with a poorer outcome of COVID-19 [57].

4.1.6. Treatment and Ventilatory Support

Previous reports emphasized that the severity of COVID-19 correlates with the risk of pulmonary fibrosis, which is also evident from our study, with all the included subjects being critically ill patients [58]. According to Zhou et al., 80% of patients infected with SARS-CoV-2 present with mild forms of the disease, 14% develop more severe symptoms, and 6% develop critical forms [59]. ICU admission is necessary for approximately 5–12%, depending on the criteria of the adopted local protocols, with most such patients being likely to benefit from ventilatory support [60,61]. Other risk factors mentioned in previous studies include ICU admission, supplemental oxygen requirement, NIMV or IMV application, and hospitalization length [53,62,63]. The patients included in the present study were critically ill, required supplemental oxygen, and benefited from NIMV or IMV during hospitalization in the ICU.

All our patients admitted to the ICU benefitted from dexamethasone therapy in accordance with the national protocol valid during the outbreak, and 19.6% received IL-6 pharmacological inhibitors. There were no statistically significant differences between patients receiving either therapeutic scheme. However, there was a correlation between a PaO_2/FiO_2 ratio of 100–200 and IL-6 receptor blockade treatment, as this treatment was initially recommended only for patients with severe forms of disease. The results of studies that evaluated the administration of corticosteroids and selective IL-6 inhibitors are few and conflicting [14]. Based on their mechanism of action, immunomodulatory agents are assumed to reduce inflammation and thus the duration of mechanical ventilation; however, it is not clear if the use of corticosteroids in the stage of ARDS reduces the risk of pulmonary fibrosis and whether the simultaneous use of corticosteroids and antifibrotics might influence long-term outcomes is currently being investigated [14]. On the contrary, in a recent meta-analysis, steroid therapy increased the risk of pulmonary fibrosis threefold [11]. McGroder et al. found that lung fibrosis was correlated with the male sex, higher SOFA scores at admission, steroid treatment, and anti-IL6 receptor blockade [15].

In 2021, Torres Castro et al. published a meta-analysis showing that 15% of COVID-19 survivors had a restrictive ventilation pattern, while 7% presented an obstructive pattern of ventilation [64]. The alteration of pulmonary functional tests can also suggest the involvement of the cardiovascular system and neuromuscular dysfunction, which can be secondary to the neurological sequelae that appear in critically ill patients [65,66].

All our patients received IMV/NIMV. Moreover, 73.21% of patients required more than 100 h of mechanical ventilation.

Duration of mechanical ventilation of over 5 days in patients with ARDS correlated with the presence of pulmonary fibrosis in 53% of ventilated patients, according to Papazian et al. [67], and the presence of fibrosis led to an increase in mortality of up to 57% [68]. It seems that, at least partially, the development of pulmonary fibrosis in ARDS can be attributed to mechanical ventilation and not only to inflammation and secondary repair processes [20,69]. The underlying pathophysiology is multifactorial, and mechanical ventilation seems to contribute to the supplementary trauma of the lungs [41], an entity known as mechanical ventilation-induced lung injury (VILI) [20,70].

The forces generated through mechanical ventilation with high transpulmonary pressure produce injury and alveolar overdistension with barotrauma, influencing the course of lung remodeling [71]. The tight junctions at the alveolar level are distorted, and the epithelial layer is interrupted and damaged, which leads to the remodeling and reorganization processes of the extracellular matrix (ECM) proteins [20,72,73]. The pulmonary ECM consists of fibronectin, elastin, collagen fibers, proteoglycans, and laminin [74], and these elements are activated by mechanical ventilation and interact with growth factors (like transforming growth factor-beta [TGF-β]), resulting in a remodeling process [75,76].

An increase in procollagen type III (PCIII) is an early marker of the fibrosis process [20,77]. Mechanical stress in isolated rat lungs was associated in experimental studies with enhanced PCIII gene expression [74,77,78]. Other studies in open-chest rabbits showed that mechanical ventilation using a high positive end-expiratory pressure is also associated with increased PCIII, procollagen type IV (PCIV) (a fibroblast growth factor), and transforming growth factor β1 (TGF-β1) gene expression [79]. In contrast, ventilation with lower pressures did not influence the expression of these genes [80].

Although we could not demonstrate the presence of VILI, 12.5% of patients in our study had no fibrosis, 67.86% presented with a minor-to-moderate degree of fibrosis, and 19.64% presented with severe fibrosis at follow-up examination.

The length of mechanical ventilation was not correlated in our group of patients with oxygen dependency at discharge or the comorbidities of the patients. However, the increased number of patients who developed pulmonary fibrosis is an additional argument for the multifactorial etiology of pulmonary fibrosis, including VILI.

4.1.7. Disease Severity

A study published in April 2021 demonstrated that the severity of the initial illness as quantified by the SOFA score, the level of LDL at admission, and the duration of mechanical ventilation are correlated with a higher risk of fibrotic-like radiographic abnormalities at 4 months after COVID-19 [15]. Moreover, the incidence of pulmonary fibrosis is 72% in those who underwent mechanical ventilation. The predicted risk of fibrotic changes varied linearly with the duration of mechanical ventilation through the first 20 days and then presented a plateau for more extended periods of mechanical ventilation [15]. In another study published in 2020 on patients admitted to an ICU in Lombardy, Italy, 88.4% (1150 out of 1300) required mechanical ventilation [81]. Out of these patients, 27 ventilated for ARDS were followed up at 110–267 days after extubation, and 23 (85%) presented pulmonary fibrosis [82], which strongly aligns with our data. Only a few studies have monitored the evolution of ARDS developed secondary to SARS-CoV-2 infection over several months in mechanically ventilated patients, but even these scarce data are helpful to provide us with a comparison term. In addition, our study reveals new information regarding the incidence of fibrosis in such patients.

4.2. Follow Up

Our study evaluated the patients 6 months after hospitalization and found an increased percentage of pulmonary fibrosis. The best time to identify irreversible pulmonary fibrosis is not well established, with some experts suggesting pulmonary functional tests and radiological investigations at 3, 6, and 12 months after an acute infectious episode [14]. Patients with pulmonary fibrosis presented lower respiratory volumes (forced vital capacity [FVC]) on pulmonary functional tests and diminished total lung capacity (TLC) [14].

The prevalence of pulmonary fibrosis in our study was 87.5%, of which 19.64% presented with severe pulmonary fibrosis. In a meta-analysis of published studies regarding pulmonary fibrosis and SARS-CoV-2 infection performed by Amin et al. in 2022, the overall prevalence of pulmonary fibrosis was found to be 44.9% [11]. Most survivors of COVID-19 suffer from postinfectious acute complications or long-term effects, especially those who present severe forms of the disease with ICU admission [14,65,83]. The most critical consequences of pulmonary fibrosis are the presence of dyspnea, limitations in physical performance (with reduced effort capacity as evidenced on pulmonary respiratory functional tests), and long-term alterations of the postinfectious quality of life. However, the good news is that in most patients, pulmonary dysfunction gradually improves over months [12]. Nabahati et al. reported that pulmonary fibrosis diminished at 6-month follow-up in 33.9% of the patients compared with those at 3-month follow-up as per thoracic CT scans [84], but patients whose fibrosis worsens over time have also been identified [11]. In 2020, Fang et al. published a study that followed 12 patients diagnosed with a severe form of COVID-19 (six treated with NIMV, four treated with IMV, and one treated with extracorporeal membrane oxygenation) and found proof of pulmonary fibrosis on CT scans performed 2 months after infection in all patients [85]. These results are also consistent with our patients' high rate of pulmonary fibrosis.

4.3. Preliminary Findings

In our study, all included patients were critically ill patients diagnosed with severe forms of ARDS, and all needed ventilatory support. Supportive oxygen therapy was used in all ventilated patients. The SOFA and APACHE scores of the monitored patients attested to the severity of the disease (as illustrated in Table 1). The treatment of these patients was challenging; critical patients had to be dealt with, among whom the mortality rate was very high, and those who survived suffered many injuries.

Fibrotic pulmonary changes seem to be a characteristic of SARS-CoV-2 infection, being more frequent in this case than in other viral cases of pneumonia (they are reported in 20% of patients after avian influenza A (H7N9) infection and 8% of patients following SARS infection) [86,87].

Pulmonary fibrosis may be a part of the long-COVID syndrome, wherein patients experience persistent symptoms and long-term consequences [3]. Patients with long-term pulmonary involvement can place an enormous burden on healthcare systems and require long-term specialty care.

4.4. General Considerations and Pathophysiological Aspects of Pulmonary Fibrosis

4.4.1. Role of Immune Cells and Cytokines

SARS-CoV-2 binds to human cells through ACE2 receptors found in many tissues such as brain, heart, lung, intestine, and kidney [12,83]. It enters through the ACE2 receptors in the pulmonary epithelium, leading to respiratory tract infection [3,14,83]. The main targets of SARS-CoV-2 in the respiratory airways are AECIIs and alveolar macrophages [88]. The virus is processed by antigen-presenting cells (APCs) and recognized by toll-like receptor 7 (TLR7) from the macrophages, enhancing intracellular signaling and activating nuclear-factor kappa B (NF-KB), with increased expression of proinflammatory cytokines [12,89,90]. APCs interact with the CD4+ cell subset of T-cells through the intermedium of major histocompatibility complex II (MHCII). CD4+ cells further proliferate and differentiate into the T helper 1 (Th1), Th2, and Th17 subclasses [91]. Th1 is responsible for the increased production of tumor necrosis factor (TNF), interferon-gamma (IFN-γ), and IL-2, while Th2 stimulates the production of IL-4, IL-5, IL-10, and IL-13, with B-cell activation and antibody production [12,92–94].

Classically, macrophages are activated via two pathways:

(1) M1 (classical)
(2) M2 (alternative)

Differentiation of the M1 phenotype is stimulated by TLR ligands and IFN-γ and is characterized by the production of proinflammatory cytokines, reactive oxygen species (ROS), reactive nitrogen, and Th1 stimulation [88]. The M1 phenotype is involved in initiating and maintaining inflammation [95]. Differentiation of the M2 phenotype is stimulated by IL-4 and IL-13 and is characterized by phagocytic activity, with a role in the inflammation resolution process [95].

Pulmonary macrophages recruited and stimulated by Th2 cytokines shift to an M2 phenotype, which also has a role in pulmonary remodeling and fibrosis in ARDS patients [20,96]. In the subsequent stages, the virus infiltrates alveolar cells (especially type 2 alveolar AECIIs), producing diffuse alveolar damage (DAD) [97], and inflammatory cytokines combine their activity with virus-induced lesions to produce lung damage [12]. The alveolar epithelial damage with the loss of epithelial barrier function produced by the virus is followed by the activation of alveolar macrophages, which initiate the process of phagocytosis of alveolar debris resulting from virus injury and release cytokines (especially IL-1 and TNF-α) and growth factors that stimulate the native cells and the regeneration of connective tissue to repair the defects [12,98,99]. The cytokine storm is the result of an immune hyperreaction due to the uncontrolled release of high amounts of cytokines that leads to multiorgan dysfunction [100], with the degree of elevation of proinflammatory cytokines correlating with the degree of disease severity [12,101–103]. The release of IL-1 and TNF-α induces the activation of adhesion molecules such as vascular adhesion molecule 1 (VCAM-1), intercellular adhesion molecule 1 (ICAM-1), and selectin, which mediate leukocyte recruitment and marginalization, rolling, and extravasation in the alveolar spaces [12]. The lesions of the alveolocapillary membrane increase the permeability of the capillary endothelium, which, in conjunction with histamine, leukotrienes, and bradykinin, allow the leakage of fluid in the interstitial and lung alveolar spaces [104]. Neutrophil infiltration in the lungs during ARDS could also modulate the fibrotic process [105,106].

4.4.2. Pulmonary Fibrosis Mechanism

Fibrosis is the consequence of an aberrant wound-healing process that follows a lung injury characterized by a distortion of normal lung architecture with lung dysfunction [12]. Fibrotic changes are secondary to inflammation, with epithelial and endothelial injury and damage to the alveolo-capillary membrane [20]. Post-COVID-19 pulmonary fibrosis is characterized by persistent fibrotic radiological/tomographic changes with functional impairment in respiratory tests at follow-up [14].

Essential components of the ECM of the basement membrane of the lungs (laminins, collagen VI, and fibronectin) are downregulated, while others are upregulated (MMP2, MMP8, and cathepsin proteins) [107]. Proinflammatory cytokines in SARS-CoV-2 infection are also believed to increase matrix metallopeptidase 1 (MMP1) and MMP7 expression, which degrades the ECM and contributes to airway remodeling and pulmonary fibrosis [14,108]. Increased CRP, IL-6, and LDH levels in the presence of systemic inflammation may activate fibroblast proliferation during the healing process of lung injury [14,109]. The proliferation of fibroblastic tissue with the excessive deposition of ECM leads to interlobular septal thickening, traction bronchiectasis with the modification of typical lung architecture, and lung fibrosis [14,110]. TGF-β is hypothesized to be directly amplified by the SARS-CoV-2 nucleocapsid protein and is also upregulated by angiotensin II, which increases in the lungs in a compensatory manner due to ACE2 receptor downregulation caused by the virus [111]. TGF-β is also produced by activated alveolar macrophages and bronchial epithelial cells and acts as a significant profibrotic stimulus [14] by stimulating the proliferation and migration of fibroblasts and activation of myofibroblasts and regulating collagen, fibronectin, and elastin formation and deposition. It also stimulates excessive ECM formation and deposition and prevents its degradation by matrix metalloproteinases, with all these factors leading to an excessive accumulation of fibrotic scar tissue [35,99,112].

Lung fibroproliferation was associated with ventilator dependency and increased mortality [113]. The pressures used in mechanical ventilation with cellular damage favor the release of inflammatory cytokines [114], accentuating lung injury and leading to inflammatory cell recruitment. Activating type 1 and type 2 helper T-cells (Th1 and Th2) causes the release of additional chemokines and growth factors, producing additional biotrauma associated with barotrauma [20,114]. ROS are produced by activated immune cells in the context of inflammation and induce alveolar epithelial apoptosis with neutrophil degranulation and oxidative stress, promoting the secretion of profibrotic cytokines with fibroblast activation [115].

Fibroblasts are mesenchymal cells found in different tissues, including the lungs. They are essential in tissue repair, secreting, and regulating the ECM [116]. Residual lung fibroblasts are activated and proliferate in response to growth factors (fibroblast growth factor [FGF], TGF-β, platelet-derived growth factor [PDGF], and epidermal growth factor [EGF]) and Th2 cytokines such as IL-1 [117] and differentiate into myofibroblasts [118]. Circulating fibrocytes originating in the bone marrow can travel from the peripheral blood to the lung and transform into interstitial fibroblasts [119,120]. The presence of fibrocytes in the broncho-alveolar lavage fluid of mechanically ventilated ARDS patients is an independent predictor of mortality [121]. Once activated, fibroblasts synthesize collagen, fibronectin, ECM components, and mediators of the repair process, such as growth factors [110]. The development of pulmonary fibrosis was correlated with increased levels of TGF-β and PCIII [20]. The growth factors vascular endothelial growth factor (VEGF) and FGF stimulate the proliferation of intact endothelial cells and the process of pulmonary capillary angiogenesis [122], while TGF-α and epidermal growth factor (EGF) stimulate the bronchiolar stem cells to regenerate the injured alveolar epithelium [123].

In our study, as observed on the CT scans, most patients presented with a minor-to-moderate degree of pulmonary fibrosis, and only 12.5% presented no fibrosis. The degree of fibrosis was not correlated with the patient's comorbidities, paraclinical examinations, or the number of hours of mechanical ventilation required.

4.5. Therapeutic Perspective

Kooistra et al. performed a study that revealed the upregulation of coagulation, inflammatory, and neutrophil extracellular trap-related pathways in patients with pulmonary fibrosis associated with severe COVID-19. They also proved that early dexamethasone treatment did not influence the severity or the incidence of pulmonary fibrosis. Instead, prednisone treatment initiated in patients with early suspicion of pulmonary fibrosis (both from the group on dexamethasone treatment and from the group that did not receive dexamethasone) was correlated with the reduction of fibrosis biomarkers and specific genes (matrix metalloproteinase 8 [MMP8], phosphodiesterase 4D [PDE4D], cysteine-rich secretory protein [CRISP3], B-cell lymphoma 2 like protein 15 [BCL2L15] that were previously upregulated towards those observed in non-pulmonary fibrosis group [124]. Other studies to prevent pulmonary fibrosis suggest that inhibitors of upregulated genes (MMP8, PDE4) in patients with pulmonary fibrosis might be a therapeutic target [124–126].

Another recent study showed that epidermic growth factor receptor (EGFR) positivity and pulmonary fibrosis are associated with increased D-dimer levels, CRP levels, and prolonged ICU stay. EGF has essential roles in cell differentiation, division, and migration and acts through its receptors (EGFR), stimulating fibroblast and vascular endothelial cell growth and proliferation. These findings can provide preliminary data for developing and using antifibrotic drugs in COVID-19 ARDS patients [127].

5. Limitations of the Study

Because critically ill patients have a low survival rate, the study included only a limited number of patients, leading to a small sample size. Additionally, it was conducted in a single medical unit and was not a multicenter study. There was also no clear distinction between pre-existing symptoms and past comorbidities with baseline testing and those caused by COVID-19 (a baseline status would be required to eliminate bias).

The evaluation of the patients was performed at 6 months; however, these patients would generally require a more extended follow-up period. Although the 56 patients regained a close-to-normal quality of life, they still presented symptoms like dyspnea and fatigue, and whether their lung lesions will disappear or persist over time requires additional studies that include such patients. More comprehensive studies or meta-analyses, including patients who underwent mechanical ventilation, are required to understand the process better.

6. Conclusions

A significant proportion of patients with COVID-19 admitted to the ICU developed pulmonary fibrosis, as observed at a 6-month post-discharge evaluation. The most important risk factors for pulmonary fibrosis identified in the medical literature were also characteristic of our patients and included severe forms of COVID-19, ARDS, oxygen supplementation, and mechanical ventilation. Such patients require dynamic monitoring, as there is a possibility of long-term pulmonary fibrosis that requires pharmacological interventions and finding new therapeutic alternatives to prevent and reduce the incidence of this complication. Radiological investigations and pulmonary functional tests should be performed to evaluate and diagnose pulmonary fibrosis in cases of SARS-CoV-2 infection, especially in severe forms.

Author Contributions: Conceptualization, methodology, writing—original draft preparation, M.S., A.S., A.R. and S.Ș.L.; software, A.R., I.C. and R.N.; formal analysis, investigation, D.C., M.S., S.R.B., A.C., D.O. and A.B.; resources, A.S. and D.F.B.; writing—review and editing, A.S., M.S., A.R. and S.Ș.L.; project administration, visualization, supervision, M.S., A.S., A.R. and A.B.; validation, all authors. All authors have read and agreed to the published version of the manuscript.

Funding: This work was supported by George Emil Palade University of Medicine, Pharmacy, Science and Technology of Targu Mures, Research Grant number 163/8/10.01.2023.

Institutional Review Board Statement: This study was conducted in accordance with the declaration of Helsinki and approved by the Ethics Committee of Târgu-Mureș County Hospital, Romania (protocol code 13942, on 28 September 2020).

Informed Consent Statement: Written informed consent was obtained from all patients included in the study.

Data Availability Statement: No new data were created or analyzed in this study. Data sharing is not applicable to this article.

Acknowledgments: This work was supported by the University of Medicine, Pharmacy, Science and Technology "George Emil Palade" of Targu Mures Research Grant number 163/8/10.01.2023. This paper was published with the support of George Emil Palade University of Medicine, Pharmacy, Sciences and Technology of Targu Mures.

Conflicts of Interest: The authors declare no conflict of interest.

Abbreviations

ACE2	angiotensin-converting enzyme 2
AECIIs	type II alveolar epithelial cells
AECIs	Type I alveolar epithelial cells
ALT	alanine transaminase
APACHE II	acute physiology and chronic health evaluation II

APCs	antigen-presenting cells
ARDS	acute respiratory distress syndrome
AST	aspartate aminotransferase
BCL2L.15	b-cell lymphoma 2 (BCL-2) like protein 15
COPD	chronic obstructive pulmonary disease
COVID-19	Corona virus disease 2019
CRISP3	cysteine-rich secretory protein 3
CRP	C-reactive protein
CT	computerized tomography
DAD	diffuse alveolar damage
ECM	extracellular matrix
EGF	epidermal growth factor
EGFR	Epidermal growth factor receptor
FGF	fibroblast growth factor
FVC	forced vital capacity
H7N9	avian influenza A
HIF-1-α	hypoxia-inducible factor 1-alpha
ICAM-1	intercellular adhesion molecule 1
ICU	intensive care unit
IFN-γ	Interferon-gamma
IMV	invasive mechanical ventilation
IPF	idiopathic pulmonary fibrosis
LDH	lactate dehydrogenase
MHCII	major histocompatibility complex II
MMP	matrix metallopeptidase
MMP-8	matrix metalloproteinase-8
NF-KB	nuclear-factor kappa B
NIMV	non-invasive mechanical ventilation
PaO_2/FiO_2	the ratio of arterial oxygen partial pressure PaO_2 to fractional inspired oxygen FiO_2
PCIII	procollagen type III
PCIV	procollagen type IV
PDE4D	phosphodiesterase 4D
PDGF	platelet-derived growth factor
qSOFA	Quick Sequential Organ Failure Assessment
ROS	Reactive oxygen species
RT-PCR	reverse transcription polymerase chain-reaction
SARS-CoV-1	Severe acute respiratory syndrome coronavirus 1
SARS-CoV-2	Severe acute respiratory syndrome coronavirus 2
SD	standard deviation
TGF-β1	transforming growth factor β
Th	T helper
TLC	total lung capacity
TLR	toll-like receptor
TNF	Tumor necrosis factor
VCAM-1	vascular adhesion molecule 1
VEGF	vascular endothelial growth factor
VILI	ventilation-induced lung injury
WBC	white blood cell

References

1. World Health Organization. Available online: https://covid19.who.int/ (accessed on 27 July 2023).
2. Wilde, H.; Mellan, T.; Hawryluk, I.; Dennis, J.M.; Denaxas, S.; Pagel, C.; Duncan, A.; Bhatt, S.; Flaxman, S.; Mateen, B.A.; et al. The association between mechanical ventilator compatible bed occupancy and mortality risk in intensive care patients with COVID-19: A national retrospective cohort study. *BMC Med.* **2021**, *19*, 213. [CrossRef]

3. Stoian, A.; Stoian, M.; Bajko, Z.; Maier, S.; Andone, S.; Cioflinc, R.A.; Motataianu, A.; Barcutean, L.; Balasa, R. Autoimmune Encephalitis in COVID-19 Infection: Our Experience and Systematic Review of the Literature. *Biomedicines* **2022**, *10*, 774. [CrossRef]
4. Stoian, A.; Bajko, Z.; Stoian, M.; Cioflinc, R.A.; Niculescu, R.; Arbănași, E.M.; Russu, E.; Botoncea, M.; Bălașa, R. The Occurrence of Acute Disseminated Encephalomyelitis in SARS-CoV-2 Infection/Vaccination: Our Experience and a Systematic Review of the Literature. *Vaccines* **2023**, *11*, 1225. [CrossRef]
5. Roman, A.; Moldovan, S.; Stoian, M.; Tilea, B.; Dobru, D. SARS-CoV-2 associated liver injury: A six-month follow-up analysis of liver function recovery. *Med. Pharm. Rep.* **2022**, *95*, 393–399. [CrossRef]
6. Boeriu, A.; Roman, A.; Dobru, D.; Stoian, M.; Voidăzan, S.; Fofiu, C. The Impact of Clostridioides Difficile Infection in Hospitalized Patients: What Changed during the Pandemic? *Diagnostics* **2022**, *12*, 3196. [CrossRef]
7. Roman, A.; Georgescu, A.M.; Moldovan, S.; Stoian, M.; Dobru, D. Liver injury in COVID-19 patients—A multidisciplinary experience and a call for national consensus. *J. Gastrointest. Liver Dis.* **2020**, *29*, 688–690. [CrossRef]
8. Roman, A.; Moldovan, S.; Santini, A.; Stoian, M.; Dobru, D. Impact of the Severity of Liver Injury in COVID-19 Patients Admitted to an Intensive Care Unit During the SARS-CoV2 Pandemic Outbreak. *J. Crit. Care Med.* **2021**, *7*, 211–216. [CrossRef]
9. He, W.; Liu, X.; Hu, B.; Li, D.; Chen, L.; Li, Y.; Tu, Y.; Xiong, S.; Wang, G.; Deng, J.; et al. Mechanisms of SARS-CoV-2 Infection-Induced Kidney Injury: A Literature Review. *Front. Cell. Infect. Microbiol.* **2022**, *12*, 838213. [CrossRef]
10. Dhakal, B.P.; Sweitzer, N.K.; Indik, J.H.; Acharya, D.; William, P. SARS-CoV-2 Infection and Cardiovascular Disease: COVID-19 Heart. *Heart Lung Circ.* **2020**, *29*, 973–987. [CrossRef]
11. Amin, B.J.H.; Kakamad, F.H.; Ahmed, G.S.; Ahmed, S.F.; Abdulla, B.A.; Mohammed, S.H.; Mikael, T.M.; Salih, R.Q.; Ali, R.K.; Salh, A.M.; et al. Post COVID-19 pulmonary fibrosis; a meta-analysis study. *Ann. Med. Surg.* **2022**, *77*, 103590. [CrossRef]
12. Ojo, A.S.; Balogun, S.A.; Williams, O.T.; Ojo, O.S. Pulmonary Fibrosis in COVID-19 Survivors: Predictive Factors and Risk Reduction Strategies. *Pulm. Med.* **2020**, *2020*, 6175964. [CrossRef] [PubMed]
13. Hui, D.S.; Wong, K.T.; Ko, F.W.; Tam, L.S.; Chan, D.P.; Woo, J.; Sung, J.J. The 1-Year Impact of Severe Acute Respiratory Syndrome on Pulmonary Function, Exercise Capacity, and Quality of Life in a Cohort of Survivors. *Chest* **2005**, *128*, 2247–2261. [CrossRef] [PubMed]
14. Duong-Quy, S.; Vo-Pham-Minh, T.; Tran-Xuan, Q.; Huynh-Anh, T.; Vo-Van, T.; Vu-Tran-Thien, Q.; Nguyen-Nhu, V. Post-COVID-19 Pulmonary Fibrosis: Facts—Challenges and Futures: A Narrative Review. *Pulm. Ther.* **2023**, *9*, 295–307. [CrossRef] [PubMed]
15. McGroder, C.F.; Zhang, D.; Choudhury, M.A.; Salvatore, M.M.; D'Souza, B.M.; Hoffman, E.A.; Wei, Y.; Baldwin, M.R.; Garcia, C.K. Pulmonary fibrosis 4 months after COVID-19 is associated with severity of illness and blood leucocyte telomere length. *Thorax* **2021**, *76*, 1242–1245. [CrossRef] [PubMed]
16. Sturgill, J.L.; Mayer, K.P.; Kalema, A.G.; Dave, K.; Mora, S.; Kalantar, A.; Carter, D.J.; Montgomery-Yates, A.A.; Morris, P.E. Post-intensive care syndrome and pulmonary fibrosis in patients surviving ARDS-pneumonia of COVID-19 and non-COVID-19 etiologies. *Sci. Rep.* **2023**, *13*, 6554. [CrossRef]
17. Kligerman, S.J.; Franks, T.J.; Galvin, J.R. From the Radiologic Pathology Archives: Organization and Fibrosis as a Response to Lung Injury in Diffuse Alveolar Damage, Organizing Pneumonia, and Acute Fibrinous and Organizing Pneumonia. *RadioGraphics* **2013**, *33*, 1951–1975. [CrossRef]
18. Tran, S.; Ksajikian, A.; Overbey, J.; Li, P.; Li, Y. Pathophysiology of Pulmonary Fibrosis in the Context of COVID-19 and Implications for Treatment: A Narrative Review. *Cells* **2022**, *11*, 2489. [CrossRef]
19. ARDS Definition of Task Force; Ranieri, V.M.; Rubenfeld, G.D.; Thompson, B.T.; Ferguson, N.D.; Caldwell, E.; Fan, E.; Camporota, L.; Slutsky, A.S. Acute Respiratory Distress Syndrome: The Berlin Definition. *JAMA* **2012**, *307*, 2526–2533. [CrossRef]
20. Cabrera-Benitez, N.E.; Laffey, J.G.; Parotto, M.; Spieth, P.M.; Villar, J.; Zhang, H.; Slutsky, A.S. Mechanical ventilation-associated lung fibrosis in acute respiratory distress syndrome: A significant contributor to poor outcome. *Anesthesiology* **2014**, *121*, 189–198. [CrossRef]
21. Michalski, J.E.; Kurche, J.S.; Schwartz, D.A. From ARDS to pulmonary fibrosis: The next phase of the COVID-19 pandemic? *Transl. Res.* **2022**, *241*, 13–24. [CrossRef]
22. Zirpe, K.G.; Tiwari, A.M.; Gurav, S.K.; Deshmukh, A.M.; Suryawanshi, P.B.; Wankhede, P.P.; Kapse, U.S.; Bhoyar, A.P.; Khan, A.Z.; Malhotra, R.V.; et al. Timing of Invasive Mechanical Ventilation and Mortality among Patients with Severe COVID-19-associated Acute Respiratory Distress Syndrome. *Indian J. Crit. Care Med.* **2021**, *25*, 493–498. [CrossRef] [PubMed]
23. Kangelaris, K.N.M.; Ware, L.B.; Wang, C.Y.M.; Janz, D.R.M.; Zhuo, H.; Matthay, M.A.; Calfee, C.S.M. Timing of Intubation and Clinical Outcomes in Adults With Acute Respiratory Distress Syndrome*. *Crit. Care Med.* **2016**, *44*, 120–129. [CrossRef]
24. Wunsch, H. Mechanical Ventilation in COVID-19: Interpreting the Current Epidemiology. *Am. J. Respir. Crit. Care Med.* **2020**, *202*, 1–4. [CrossRef] [PubMed]
25. Wang, Y.; Lu, X.; Li, Y.; Chen, H.; Chen, T.; Su, N.; Huang, F.; Zhou, J.; Zhang, B.; Yan, F.; et al. Clinical Course and Outcomes of 344 Intensive Care Patients with COVID-19. *Am. J. Respir. Crit. Care Med.* **2020**, *201*, 1430–1434. [CrossRef]
26. Udi, J.; Lang, C.N.; Zotzmann, V.; Krueger, K.; Fluegler, A.; Bamberg, F.; Bode, C.; Duerschmied, D.; Wengenmayer, T.; Staudacher, D.L. Incidence of Barotrauma in Patients With COVID-19 Pneumonia During Prolonged Invasive Mechanical Ventilation—A Case-Control Study. *J. Intensiv. Care Med.* **2020**, *36*, 477–483. [CrossRef] [PubMed]
27. Sirayder, U.; Inal-Ince, D.; Kepenek-Varol, B.; Acik, C. Long-Term Characteristics of Severe COVID-19: Respiratory Function, Functional Capacity, and Quality of Life. *Int. J. Environ. Res. Public Health* **2022**, *19*, 6304. [CrossRef]

28. Seymour, C.W.; Liu, V.X.; Iwashyna, T.J.; Brunkhorst, F.M.; Rea, T.D.; Scherag, A.; Rubenfeld, G.; Kahn, J.M.; Shankar-Hari, M.; Singer, M.; et al. Assessment of Clinical Criteria for Sepsis: For the Third International Consensus Definitions for Sepsis and Septic Shock (Sepsis-3). *JAMA* **2016**, *315*, 762–774. [CrossRef]
29. Knaus, W.A.; Draper, E.A.; Wagner, D.P.; Zimmerman, J.E. APACHE II: A severity of disease classification system. *Crit. Care Med.* **1985**, *13*, 818–829. [CrossRef]
30. Barkauskas, C.E.; Cronce, M.J.; Rackley, C.R.; Bowie, E.J.; Keene, D.R.; Stripp, B.R.; Randell, S.H.; Noble, P.W.; Hogan, B.L.M. Type 2 alveolar cells are stem cells in adult lung. *J. Clin. Investig.* **2013**, *123*, 3025–3036. [CrossRef]
31. Zou, X.; Chen, K.; Zou, J.; Han, P.; Hao, J.; Han, Z. Single-cell RNA-seq data analysis on the receptor ACE2 expression reveals the potential risk of different human organs vulnerable to 2019-nCoV infection. *Front. Med.* **2020**, *14*, 185–192. [CrossRef]
32. Hamming, I.; Timens, W.; Bulthuis, M.L.C.; Lely, A.T.; Navis, G.J.; van Goor, H. Tissue distribution of ACE2 protein, the functional receptor for SARS coronavirus. A first step in understanding SARS pathogenesis. *J. Pathol.* **2004**, *203*, 631–637. [CrossRef] [PubMed]
33. Sungnak, W.; Huang, N.; Becavin, C.; Berg, M.; Queen, R.; Litvinukova, M.; Talavera-Lopez, C.; Maatz, H.; Reichart, D.; Sampaziotis, F.; et al. HCA Lung Biological Network. SARS-CoV-2 entry factors are highly expressed in nasal epithelial cells together with innate immune genes. *Nat. Med.* **2020**, *26*, 681–687. [CrossRef] [PubMed]
34. Kakamad, F.H.; Mahmood, S.O.; Rahim, H.M.; Abdulla, B.A.; Abdullah, H.O.; Othman, S.; Mohammed, S.H.; Kakamad, S.H.; Mustafa, S.M.; Salih, A.M. Post covid-19 invasive pulmonary Aspergillosis: A case report. *Int. J. Surg. Case Rep.* **2021**, *82*, 105865. [CrossRef] [PubMed]
35. Venkataraman, T.; Frieman, M.B. The role of epidermal growth factor receptor (EGFR) signaling in SARS coronavirus-induced pulmonary fibrosis. *Antivir. Res.* **2017**, *143*, 142–150. [CrossRef]
36. Mo, X.; Jian, W.; Su, Z.; Chen, M.; Peng, H.; Peng, P.; Lei, C.; Chen, R.; Zhong, N.; Li, S. Abnormal pulmonary function in COVID-19 patients at time of hospital discharge. *Eur. Respir. J.* **2020**, *55*, 2001217. [CrossRef]
37. Frija-Masson, J.; Debray, M.-P.; Gilbert, M.; Lescure, F.-X.; Travert, F.; Borie, R.; Khalil, A.; Crestani, B.; D'Ortho, M.-P.; Bancal, C. Functional characteristics of patients with SARS-CoV-2 pneumonia at 30 days post-infection. *Eur. Respir. J.* **2020**, *56*, 2001754. [CrossRef]
38. Stoian, A.; Motataianu, A.; Bajko, Z.; Balasa, A. Guillain–Barré and Acute Transverse Myelitis Overlap Syndrome Following Obstetric Surgery. *J. Crit. Care Med.* **2020**, *6*, 74–79. [CrossRef]
39. Ahmed, O.F.; Kakamad, F.H.; Amin, B.J.H.; Abdullah, B.A.; Hassan, M.N.; Salih, R.Q.; Mohammed, S.H.; Othman, S.; Ahmed, G.S.; Salih, A.M. Post COVID-19 pulmonary complications; a single center experience. *Ann. Med. Surg.* **2021**, *72*, 103052. [CrossRef]
40. Antonio, G.E.; Wong, K.T.; Hui, D.S.C.; Wu, A.; Lee, N.; Yuen, E.H.Y.; Leung, C.B.; Rainer, T.H.; Cameron, P.; Chung, S.S.C.; et al. Thin-Section CT in Patients with Severe Acute Respiratory Syndrome Following Hospital Discharge: Preliminary Experience. *Radiology* **2003**, *228*, 810–815. [CrossRef]
41. Ambardar, S.R.; Hightower, S.L.; Huprikar, N.A.; Chung, K.K.; Singhal, A.; Collen, J.F. Post-COVID-19 Pulmonary Fibrosis: Novel Sequelae of the Current Pandemic. *J. Clin. Med.* **2021**, *10*, 2452. [CrossRef]
42. Crooks, M.G.; Hart, S.P. Coagulation and anticoagulation in idiopathic pulmonary fibrosis. *Eur. Respir. Rev.* **2015**, *24*, 392–399. [CrossRef] [PubMed]
43. Aquino-Gálvez, A.; González-Ávila, G.; Jiménez-Sánchez, L.L.; Maldonado-Martínez, H.A.; Cisneros, J.; Toscano-Marquez, F.; Castillejos-López, M.; Torres-Espíndola, L.M.; Velázquez-Cruz, R.; Rodríguez, V.H.O.; et al. Dysregulated expression of hypoxia-inducible factors augments myofibroblasts differentiation in idiopathic pulmonary fibrosis. *Respir. Res.* **2019**, *20*, 130. [CrossRef] [PubMed]
44. Chen, I.-T.; Huang, L.-T.; Chen, C.-C.; Chen, C.-M. Molecular mechanisms underlying hyperoxia-induced lung fibrosis. *Pediatr. Neonatol.* **2022**, *63*, 109–116. [CrossRef] [PubMed]
45. Mach, W.J.; Thimmesch, A.R.; Pierce, J.T.; Pierce, J.D. Consequences of Hyperoxia and the Toxicity of Oxygen in the Lung. *Nurs. Res. Pr.* **2011**, *2011*, 260482. [CrossRef]
46. Chernyak, B.V.; Popova, E.N.; Prikhodko, A.S.; Grebenchikov, O.A.; Zinovkina, L.A.; Zinovkin, R.A. COVID-19 and Oxidative Stress. *Biochemistry* **2020**, *85*, 1543–1553. [CrossRef]
47. Fukumoto, J.; Leung, J.; Cox, R.; Czachor, A.; Parthasarathy, P.T.; Lagishetty, V.; Mandry, M.; Hosseinian, N.; Patel, P.; Perry, B.; et al. Oxidative stress induces club cell proliferation and pulmonary fibrosis in Atp8b1 mutant mice. *Aging* **2019**, *11*, 209–229. [CrossRef]
48. Voshaar, T.; Stais, P.; Köhler, D.; Dellweg, D. Conservative management of COVID-19 associated hypoxaemia. *ERJ Open Res.* **2021**, *7*, 00026–02021. [CrossRef]
49. Vianello, A.; Guarnieri, G.; Braccioni, F.; Lococo, S.; Molena, B.; Cecchetto, A.; Giraudo, C.; De Marchi, L.B.; Caminati, M.; Senna, G. The pathogenesis, epidemiology and biomarkers of susceptibility of pulmonary fibrosis in COVID-19 survivors. *Clin. Chem. Lab. Med.* **2021**, *60*, 307–316. [CrossRef]
50. John, A.E.; Joseph, C.; Jenkins, G.; Tatler, A.L. COVID-19 and pulmonary fibrosis: A potential role for lung epithelial cells and fibroblasts. *Immunol. Rev.* **2021**, *302*, 228–240. [CrossRef]
51. Xu, J.; Gonzalez, E.T.; Iyer, S.S.; Mac, V.; Mora, A.L.; Sutliff, R.L.; Reed, A.; Brigham, K.L.; Kelly, P.; Rojas, M. Use of Senescence-Accelerated Mouse Model in Bleomycin-Induced Lung Injury Suggests That Bone Marrow-Derived Cells Can Alter the Outcome of Lung Injury in Aged Mice. *J. Gerontol. A Biol. Sci. Med. Sci.* **2009**, *64*, 731–739. [CrossRef]

52. Cilli, A.; Hanta, I.; Uzer, F.; Coskun, F.; Sevinc, C.; Deniz, P.P.; Parlak, M.; Altunok, E.; Tertemiz, K.C.; Ursavas, A. Characteristics and outcomes of COVID-19 patients with IPF: A multi-center retrospective study. *Respir. Med. Res.* **2022**, *81*, 100900. [CrossRef] [PubMed]
53. Li, X.; Shen, C.; Wang, L.; Majumder, S.; Zhang, D.; Deen, M.J.; Li, Y.; Qing, L.; Zhang, Y.; Chen, C.; et al. Pulmonary fibrosis and its related factors in discharged patients with new corona virus pneumonia: A cohort study. *Respir. Res.* **2021**, *22*, 203. [CrossRef] [PubMed]
54. Huang, C.; Huang, L.; Wang, Y.; Li, X.; Ren, L.; Gu, X.; Kang, L.; Guo, L.; Liu, M.; Zhou, X.; et al. RETRACTED: 6-month consequences of COVID-19 in patients discharged from hospital: A cohort study. *Lancet* **2021**, *397*, 220–232. [CrossRef]
55. Nalbandian, A.; Sehgal, K.; Gupta, A.; Madhavan, M.V.; McGroder, C.; Stevens, J.S.; Cook, J.R.; Nordvig, A.S.; Shalev, D.; Sehrawat, T.S.; et al. Post-acute COVID-19 syndrome. *Nat. Med.* **2021**, *27*, 601–615. [CrossRef]
56. Rajagopal, K.; Bryant, A.J.; Sahay, S.; Wareing, N.; Zhou, Y.; Pandit, L.M.; Karmouty-Quintana, H. Idiopathic pulmonary fibrosis and pulmonary hypertension: Heracles meets the Hydra. *Br. J. Pharmacol.* **2021**, *178*, 172–186. [CrossRef]
57. Jdiaa, S.S.; Mansour, R.; El Alayli, A.; Gautam, A.; Thomas, P.; Mustafa, R.A. COVID-19 and chronic kidney disease: An updated overview of reviews. *J. Nephrol.* **2022**, *35*, 69–85. [CrossRef] [PubMed]
58. Zou, J.-N.; Sun, L.; Wang, B.-R.; Zou, Y.; Xu, S.; Ding, Y.-J.; Shen, L.-J.; Huang, W.-C.; Jiang, X.-J.; Chen, S.-M. The characteristics and evolution of pulmonary fibrosis in COVID-19 patients as assessed by AI-assisted chest HRCT. *PLoS ONE* **2021**, *16*, e0248957. [CrossRef]
59. Zhou, F.; Yu, T.; Du, R.; Fan, G.; Liu, Y.; Liu, Z.; Xiang, J.; Wang, Y.; Song, B.; Gu, X.; et al. Clinical course and risk factors for mortality of adult inpatients with COVID-19 in Wuhan, China: A retrospective cohort study. *Lancet* **2020**, *395*, 1054–1062. [CrossRef]
60. Grasselli, G.; Pesenti, A.; Cecconi, M. Critical Care Utilization for the COVID-19 Outbreak in Lombardy, Italy: Early experience and forecast during an emergency response. *JAMA* **2020**, *323*, 1545–1546. [CrossRef]
61. Guan, W.J.; Ni, Z.Y.; Hu, Y.; Liang, W.H.; Ou, C.Q.; He, J.X.; Liu, L.; Shan, H.; Lei, C.L.; Hui, D.S.C.; et al. China Medical Treatment Expert Group for Covid-19. Clinical Characteristics of Coronavirus Disease 2019 in China. *N. Engl. J. Med.* **2020**, *382*, 1708–1720. [CrossRef]
62. Aul, R.; Gates, J.; Draper, A.; Dunleavy, A.; Ruickbie, S.; Meredith, H.; Walters, N.; van Zeller, C.; Taylor, V.; Bridgett, M.; et al. Complications after discharge with COVID-19 infection and risk factors associated with development of post-COVID pulmonary fibrosis. *Respir. Med.* **2021**, *188*, 106602. [CrossRef] [PubMed]
63. Yu, M.; Liu, Y.; Xu, D.; Zhang, R.; Lan, L.; Xu, H. Prediction of the Development of Pulmonary Fibrosis Using Serial Thin-Section CT and Clinical Features in Patients Discharged after Treatment for COVID-19 Pneumonia. *Korean J. Radiol.* **2020**, *21*, 746–755. [CrossRef] [PubMed]
64. Torres-Castro, R.; Vasconcello-Castillo, L.; Alsina-Restoy, X.; Solis-Navarro, L.; Burgos, F.; Puppo, H.; Vilaró, J. Respiratory function in patients post-infection by COVID-19: A systematic review and meta-analysis. *Pulmonology* **2020**, *27*, 328–337. [CrossRef] [PubMed]
65. Stoian, A.; Bajko, Z.; Maier, S.; Cioflinc, R.A.; Grigorescu, B.L.; Moțățăianu, A.; Bărcuțean, L.; Balașa, R.; Stoian, M. High-dose intravenous immunoglobulins as a therapeutic option in critical illness polyneuropathy accompanying SARS-CoV-2 infection: A case-based review of the literature (Review). *Exp. Ther. Med.* **2021**, *22*, 1182. [CrossRef] [PubMed]
66. Stoian, A.; Moțățăianu, A.; Bărcuțean, L.; Maier, S.; Bajko, Z.; Voidăzan, S.; Fărcaș, A.; Bălașa, R. Understandig the mechanism of action of intravenous immunoglobulins: A ten years' experience in treating Guillain Barrésyndrome. *Farmacia* **2020**, *68*, 426–435. [CrossRef]
67. Papazian, L.; Doddoli, C.; Chetaille, B.; Gernez, Y.; Thirion, X.; Roch, A.; Donati, Y.; Bonnety, M.; Zandotti, C.; Thomas, P. A contributive result of open-lung biopsy improves survival in acute respiratory distress syndrome patients. *Crit. Care Med.* **2007**, *35*, 755–762. [CrossRef]
68. Madtes, D.K.; Rubenfeld, G.; Klima, L.D.; Milberg, J.A.; Steinberg, K.P.; Martin, T.R.; Raghu, G.; Hudson, L.D.; Clark, J.G. Elevated Transforming Growth Factor- alpha Levels in Bronchoalveolar Lavage Fluid of Patients with Acute Respiratory Distress Syndrome. *Am. J. Respir. Crit. Care Med.* **1998**, *158*, 424–430. [CrossRef]
69. Martin, C.; Papazian, L.; Payan, M.-J.; Saux, P.; Gouin, F. Pulmonary fibrosis correlates with outcome in adult respiratory distress syndrome. A study in mechanically ventilated patients. *Chest* **1995**, *107*, 196–200. [CrossRef]
70. Li, H.-H.; Wang, C.-W.; Chang, C.-H.; Huang, C.-C.; Hsu, H.-S.; Chiu, L.-C. Relationship between Mechanical Ventilation and Histological Fibrosis in Patients with Acute Respiratory Distress Syndrome Undergoing Open Lung Biopsy. *J. Pers. Med.* **2022**, *12*, 474. [CrossRef]
71. Copland, I.B.; Reynaud, D.; Pace-Asciak, C.; Post, M. Mechanotransduction of stretch-induced prostanoid release by fetal lung epithelial cells. *Am. J. Physiol. Lung Cell Mol. Physiol.* **2006**, *291*, L487–L495. [CrossRef]
72. Cavanaugh, K.J., Jr.; Oswari, J.; Margulies, S.S. Role of Stretch on Tight Junction Structure in Alveolar Epithelial Cells. *Am. J. Respir. Cell Mol. Biol.* **2001**, *25*, 584–591. [CrossRef]
73. Heise, R.L.; Stober, V.; Cheluvaraju, C.; Hollingsworth, J.W.; Garantziotis, S. Mechanical Stretch Induces Epithelial-Mesenchymal Transition in Alveolar Epithelia via Hyaluronan Activation of Innate Immunity. *J. Biol. Chem.* **2011**, *286*, 17435–17444. [CrossRef] [PubMed]

74. Parker, J.C.; Breen, E.C.; West, J.B.; Maina, J.N.; Farias, L.L.; Faffe, D.S.; Xisto, D.G.; Santana, M.C.E.; Lassance, R.; Prota, L.F.M.; et al. High vascular and airway pressures increase interstitial protein mRNA expression in isolated rat lungs. *J. Appl. Physiol.* **1997**, *83*, 1697–1705. [CrossRef]
75. Tschumperlin, D.J.; Dai, G.; Maly, I.V.; Kikuchi, T.; Laiho, L.H.; McVittie, A.K.; Haley, K.J.; Lilly, C.M.; So, P.T.C.; Lauffenburger, D.A.; et al. Mechanotransduction through growth-factor shedding into the extracellular space. *Nature* **2004**, *429*, 83–86. [CrossRef] [PubMed]
76. Schild, C.; Trueb, B. Mechanical Stress Is Required for High-Level Expression of Connective Tissue Growth Factor. *Exp. Cell Res.* **2002**, *274*, 83–91. [CrossRef] [PubMed]
77. Garcia, C.S.; Rocco, P.R.; Facchinetti, L.D.; Lassance, R.M.; Caruso, P.; Deheinzelin, D.; Morales, M.M.; Romero, P.V.; Faffe, D.S.; Zin, W.A. What increases type III procollagen mRNA levels in lung tissue: Stress induced by changes in force or amplitude? *Respir. Physiol. Neurobiol.* **2004**, *144*, 59–70. [CrossRef] [PubMed]
78. Cruz, F.F.; Rocco, P.R.M.; Pelosi, P. Role of the extracellular matrix in the genesis of ventilator-induced lung injury. *Med. Klin. Intensiv. Notfmed.* **2018**, *113* (Suppl. 1), 2–6. [CrossRef]
79. Tatler, A.L.; Jenkins, G. TGF-β Activation and Lung Fibrosis. *Proc. Am. Thorac. Soc.* **2012**, *9*, 130–136. [CrossRef] [PubMed]
80. Berg, J.T.; Fu, Z.; Breen, E.C.; Tran, H.-C.; Mathieu-Costello, O.; West, J.B.; Leuenberger, A.; Gazdhar, A.; Herrmann, G.; Ochs, M.; et al. High lung inflation increases mRNA levels of ECM components and growth factors in lung parenchyma. *J. Appl. Physiol.* **1997**, *83*, 120–128. [CrossRef]
81. Grasselli, G.; Zangrillo, A.; Zanella, A.; Antonelli, M.; Cabrini, L.; Castelli, A.; Cereda, D.; Coluccello, A.; Foti, G.; Fumagalli, R.; et al. COVID-19 Lombardy ICU Network. Baseline Characteristics and Outcomes of 1591 Patients Infected With SARS-CoV-2 Admitted to ICUs of the Lombardy Region, Italy. *JAMA* **2020**, *323*, 1574–1581, Erratum in *JAMA* **2021**, *325*, 2120. [CrossRef]
82. Desai, S.R.; Wells, A.U.; Rubens, M.B.; Evans, T.W.; Hansell, D.M. Acute Respiratory Distress Syndrome: CT Abnormalities at Long-term Follow-up. *Radiology* **1999**, *210*, 29–35. [CrossRef] [PubMed]
83. Stoian, A.; Bălașa, R.; Grigorescu, B.L.; Maier, S.; Andone, S.; Cocuz, I.G.; Bajko, Z.; Filep, C.R.; Stoian, M. Guillain-Barré syndrome associated with Covid-19: A close relationship or just a coincidence? (Review). *Exp. Ther. Med.* **2021**, *22*, 916. [CrossRef]
84. Nabahati, M.; Ebrahimpour, S.; Khaleghnejad Tabari, R.; Mehraeen, R. Post-COVID-19 pulmonary fibrosis and its predictive factors: A prospective study. *Egypt J. Radiol. Nucl. Med.* **2021**, *52*, 248. [CrossRef]
85. Fang, Y.; Zhou, J.; Ding, X.; Ling, G.; Yu, S. Pulmonary fibrosis in critical ill patients recovered from COVID-19 pneumonia: Preliminary experience. *Am. J. Emerg. Med.* **2020**, *38*, 2134–2138. [CrossRef] [PubMed]
86. Ali, R.M.M.; Ghonimy, M.B.I. Post-COVID-19 pneumonia lung fibrosis: A worrisome sequelae in surviving patients. *Egypt J. Radiol. Nucl. Med.* **2021**, *52*, 101. [CrossRef]
87. Udwadia, Z.; Pokhariyal, P.; Tripathi, A.R.; Kohli, A. Fibrotic interstitial lung disease occurring as sequelae of COVID-19 pneumonia despite concomitant steroids. *Lung India* **2021**, *38* (Suppl. 1). [CrossRef] [PubMed]
88. Patrucco, F.; Solidoro, P.; Gavelli, F.; Apostolo, D.; Bellan, M. Idiopathic Pulmonary Fibrosis and Post-COVID-19 Lung Fibrosis: Links and Risks. *Microorganisms* **2023**, *11*, 895. [CrossRef] [PubMed]
89. Zumla, A.; Hui, D.S.; Azhar, E.I.; Memish, Z.A.; Maeurer, M. Reducing mortality from 2019-nCoV: Host-directed therapies should be an option. *Lancet* **2020**, *395*, e35–e36. [CrossRef]
90. Strieter, R.M.; Mehrad, B. New Mechanisms of Pulmonary Fibrosis. *Chest* **2009**, *136*, 1364–1370. [CrossRef]
91. Wang, J.; Wang, B.J.; Yang, J.C.; Wang, M.Y.; Chen, C.; Luo, G.X.; He, W.F. Research advances in the mechanism of pulmonary fibrosis induced by coronavirus disease 2019 and the corresponding therapeutic measures. *Zhonghua Shao Shang Za Zhi* **2020**, *36*, 691–697. (In Chinese) [CrossRef]
92. Romagnani, S. Th1/Th2 Cells. *Inflamm. Bowel Dis.* **1999**, *5*, 285–294. [CrossRef] [PubMed]
93. Guglani, L.; A Khader, S. Th17 cytokines in mucosal immunity and inflammation. *Curr. Opin. HIV AIDS* **2010**, *5*, 120–127. [CrossRef] [PubMed]
94. Liu, W.; Fontanet, A.; Zhang, P.-H.; Zhan, L.; Xin, Z.-T.; Baril, L.; Tang, F.; Lv, H.; Cao, W.-C. Two-Year Prospective Study of the Humoral Immune Response of Patients with Severe Acute Respiratory Syndrome. *J. Infect. Dis.* **2006**, *193*, 792–795. [CrossRef] [PubMed]
95. Martinez, F.O.; Helming, L.; Gordon, S. Alternative Activation of Macrophages: An Immunologic Functional Perspective. *Annu. Rev. Immunol.* **2009**, *27*, 451–483. [CrossRef] [PubMed]
96. Pechkovsky, D.V.; Prasse, A.; Kollert, F.; Engel, K.M.; Dentler, J.; Luttmann, W.; Friedrich, K.; Müller-Quernheim, J.; Zissel, G. Alternatively activated alveolar macrophages in pulmonary fibrosis—Mediator production and intracellular signal transduction. *Clin. Immunol.* **2010**, *137*, 89–101. [CrossRef]
97. Mohammadi, A.; Balan, I.; Yadav, S.; Matos, W.F.; Kharawala, A.; Gaddam, M.; Sarabia, N.; Koneru, S.C.; Suddapalli, S.K.; Marzban, S. Post-COVID-19 Pulmonary Fibrosis. *Cureus* **2022**, *14*, e22770. [CrossRef]
98. Huang, W.-T.; Akhter, H.; Jiang, C.; MacEwen, M.; Ding, Q.; Antony, V.; Thannickal, V.J.; Liu, R.-M. Plasminogen activator inhibitor 1, fibroblast apoptosis resistance, and aging-related susceptibility to lung fibrosis. *Exp. Gerontol.* **2015**, *61*, 62–75. [CrossRef]
99. Venkataraman, T.; Coleman, C.M.; Frieman, M.B. Overactive epidermal growth factor receptor signaling leads to increased fibrosis after severe acute respiratory syndrome coronavirus infection. *J. Virol.* **2017**, *91*, 10–1128. [CrossRef]

100. Mehta, P.; McAuley, D.F.; Brown, M.; Sanchez, E.; Tattersall, R.S.; Manson, J.J.; on behalf of theHLH Across Speciality Collaboration, UK. COVID-19: Consider cytokine storm syndromes and immunosuppression. *Lancet* **2020**, *395*, 1033–1034. [CrossRef]
101. Huang, C.; Wang, Y.; Li, X.; Ren, L.; Zhao, J.; Hu, Y.; Zhang, L.; Fan, G.; Xu, J.; Gu, X.; et al. Clinical features of patients infected with 2019 novel coronavirus in Wuhan, China. *Lancet* **2020**, *395*, 497–506. [CrossRef]
102. Balasa, R.; Maier, S.; Barcutean, L.; Stoian, A.; Motataianu, A. The direct deleterious effect of Th17 cells in the nervous system compartment in multiple sclerosis and experimental autoimmune encephalomyelitis: One possible link between neuroinflammation and neurodegeneration. *Rev. Rom. Med. Lab.* **2020**, *28*, 9–17. [CrossRef]
103. Maier, S.; Moțățăianu, A.; Bărcuțean, L.; Balint, A.; Huțanu, A.; Bajko, Z.; Stoian, A.; Andone, S.; Bălașa, R. Interferon β 1A, an immunomodulator in relapsing remitting multiple sclerosis patients: The effect on pro inflammatory cytokines. *Farmacia* **2020**, *68*, 65–75. [CrossRef]
104. Wilson, M.S.; Wynn, T.A. Pulmonary fibrosis: Pathogenesis, etiology and regulation. *Mucosal Immunol.* **2009**, *2*, 103–121. [CrossRef] [PubMed]
105. Grommes, J.; Soehnlein, O. Contribution of Neutrophils to Acute Lung Injury. *Mol. Med.* **2010**, *17*, 293–307. [CrossRef]
106. Malech, H.L.; Gallin, J.I. Current concepts: Immunology. Neutrophils in human diseases. *N. Engl. J. Med.* **1987**, *317*, 687–694. [CrossRef] [PubMed]
107. Leng, L.; Cao, R.; Ma, J.; Mou, D.; Zhu, Y.; Li, W.; Lv, L.; Gao, D.; Zhang, S.; Gong, F.; et al. Pathological features of COVID-19-associated lung injury: A preliminary proteomics report based on clinical samples. *Signal Transduct. Target. Ther.* **2020**, *5*, 240. [CrossRef] [PubMed]
108. Inoue, Y.; Kaner, R.J.; Guiot, J.; Maher, T.M.; Tomassetti, S.; Moiseev, S.; Kuwana, M.; Brown, K.K. Diagnostic and Prognostic Biomarkers for Chronic Fibrosing Interstitial Lung Diseases with a Progressive Phenotype. *Chest* **2020**, *158*, 646–659. [CrossRef]
109. Tanni, S.E.; Fabro, A.T.; de Albuquerque, A.; Ferreira, E.V.M.; Verrastro, C.G.Y.; Sawamura, M.V.Y.; Ribeiro, S.M.; Baldi, B.G. Pulmonary fibrosis secondary to COVID-19: A narrative review. *Expert Rev. Respir. Med.* **2021**, *15*, 791–803. [CrossRef]
110. Sgalla, G.; Iovene, B.; Calvello, M.; Ori, M.; Varone, F.; Richeldi, L. Idiopathic pulmonary fibrosis: Pathogenesis and management. *Respir. Res.* **2018**, *19*, 32. [CrossRef]
111. Gentile, F.; Aimo, A.; Forfori, F.; Catapano, G.; Clemente, A.; Cademartiri, F.; Emdin, M.; Giannoni, A. COVID-19 and risk of pulmonary fibrosis: The importance of planning ahead. *Eur. J. Prev. Cardiol.* **2020**, *27*, 1442–1446. [CrossRef]
112. Shukla, A.; Meisler, N.; Cutroneo, K.R. Perspective article: Transforming growth factor-beta: Crossroad of glucocorticoid and bleomycin regulation of collagen synthesis in lung fibroblasts. *Wound Repair. Regen* **1999**, *7*, 133–140. [CrossRef]
113. Ichikado, K.; Muranaka, H.; Gushima, Y.; Kotani, T.; Nader, H.M.; Fujimoto, K.; Johkoh, T.; Iwamoto, N.; Kawamura, K.; Nagano, J.; et al. Fibroproliferative changes on high-resolution CT in the acute respiratory distress syndrome predict mortality and ventilator dependency: A prospective observational cohort study. *BMJ Open* **2012**, *2*, e000545. [CrossRef] [PubMed]
114. Tremblay, L.N.; Slutsky, A.S. Ventilator-induced injury: From barotrauma to biotrauma. *Proc. Assoc. Am. Physicians* **1998**, *110*, 482–488. [PubMed]
115. Cheresh, P.; Kim, S.-J.; Tulasiram, S.; Kamp, D.W. Oxidative stress and pulmonary fibrosis. *Biochim. Biophys. Acta (BBA)—Mol. Basis Dis.* **2013**, *1832*, 1028–1040. [CrossRef]
116. Baum, J.B.; Duffy, H.S. Fibroblasts and Myofibroblasts: What Are We Talking About? *J. Cardiovasc. Pharmacol.* **2011**, *57*, 376–379. [CrossRef]
117. Phan, S. Fibroblast phenotypes in pulmonary fibrosis. *Am. J. Respir. Cell Mol. Biol.* **2003**, *29* (Suppl. 3), S87–S92. [PubMed]
118. Basil, M.H.; Hantash, B.M.; Zhao, L.; A Knowles, J.; Lorenz, H.P. Adult and fetal wound healing. *Front. Biosci.* **2008**, *13*, 51–61. [CrossRef]
119. Hashimoto, N.; Jin, H.; Liu, T.; Chensue, S.W.; Phan, S.H. Bone marrow–derived progenitor cells in pulmonary fibrosis. *J. Clin. Investig.* **2004**, *113*, 243–252. [CrossRef]
120. Phillips, R.J.; Burdick, M.D.; Hong, K.; Lutz, M.A.; Murray, L.A.; Xue, Y.Y.; Belperio, J.A.; Keane, M.P.; Strieter, R.M. Circulating fibrocytes traffic to the lungs in response to CXCL12 and mediate fibrosis. *J. Clin. Investig.* **2004**, *114*, 438–446. [CrossRef]
121. Quesnel, C.; Piednoir, P.; Gelly, J.; Nardelli, L.; Garnier, M.; Leçon, V.; Lasocki, S.; Bouadma, L.; Philip, I.; Elbim, C.; et al. Alveolar fibrocyte percentage is an independent predictor of poor outcome in patients with acute lung injury*. *Crit. Care Med.* **2012**, *40*, 21–28. [CrossRef]
122. Barrientos, S.; Stojadinovic, O.; Golinko, M.S.; Brem, H.; Tomic-Canic, M. Growth factors and cytokines in wound healing. *Wound Repair. Regen* **2008**, *16*, 585–601. [CrossRef]
123. Raja, R.; Sivamani, K.; Garcia, M.S.; Isseroff, R.R. Wound re-epithelialization: Modulating kerationcyte migration in wound healing. *Front. Biosci.* **2007**, *12*, 2849–2868. [CrossRef] [PubMed]
124. Kooistra, E.J.; Dahm, K.; van Herwaarden, A.E.; Gerretsen, J.; Nuesch Germano, M.; Mauer, K.; Smeets, R.L.; van der Velde, S.; van den Berg, M.J.W.; van der Hoeven, J.G.; et al. Molecular mechanisms and treatment responses of pulmonary fibrosis in severe COVID-19. *Respir. Res.* **2023**, *24*, 196. [CrossRef] [PubMed]
125. Craig, V.J.; Quintero, P.A.; Fyfe, S.E.; Patel, A.S.; Knolle, M.D.; Kobzik, L.; Owen, C.A. Profibrotic Activities for Matrix Metalloproteinase-8 during Bleomycin-Mediated Lung Injury. *J. Immunol.* **2013**, *190*, 4283–4296. [CrossRef] [PubMed]

126. Li, H.; Zuo, J.; Tang, W. Phosphodiesterase-4 Inhibitors for the Treatment of Inflammatory Diseases. *Front. Pharmacol.* **2018**, *9*, 1048. [CrossRef] [PubMed]
127. Dülger, S.U.; Mutlu, N.; Ceylan, I.; Özhan, E. The relationship between lung fibrosis, the epidermal growth factor receptor, and disease outcomes in COVID-19 pneumonia: A postmortem evaluation. *Clin. Exp. Med.* **2023**, *23*, 1181–1188. [CrossRef]

biomedicines

Article

Risk of Acute Myocardial Infarction in Pneumoconiosis: Results from a Retrospective Cohort Study

Ju-Hsin Chang [1,2,†], Te-Chun Shen [3,4,5,†], Ke-Wei Chen [1,6], Cheng-Li Lin [7], Chung Y. Hsu [7], Yeong-Ray Wen [1,2,*] and Kuan-Cheng Chang [4,6,*]

1 Graduate Institute of Clinical Medicine Science, China Medical University, No. 91, Xue-Shi Road, Taichung 404, Taiwan
2 Department of Anesthesia, China Medical University Hospital, Taichung 404, Taiwan
3 Division of Pulmonary and Critical Care Medicine, Department of Internal Medicine, China Medical University Hospital, Taichung 404, Taiwan
4 School of Medicine, China Medical University, Taichung 404, Taiwan
5 Department of Critical Care Medicine, Chu Shang Show Chwan Hospital, Nantou 557, Taiwan
6 Division of Cardiology, Department of Internal Medicine, China Medical University Hospital, No. 2 Yu-De Road, Taichung 404, Taiwan
7 Management Office for Health Data, China Medical University Hospital, Taichung 404, Taiwan
* Correspondence: d95236@mail.cmuh.org.tw (Y.-R.W.); d5267@mail.cmuh.org.tw (K.-C.C.); Tel.: +886-4-2205-3366 (Y.-R.W.); +886-4-2205-2121 (K.-C.C.)
† These authors contributed equally to this work.

Abstract: Background: Pneumoconiosis (PCN) has several comorbidities, most notably pulmonary and cardiovascular diseases. However, much is still unknown about the relationship between PCN and acute myocardial infarction (AMI). The present study aimed to clarify the association between PCN and subsequent AMI risk using a retrospective cohort study design. Methods: This was a population-based, retrospective cohort study that used data from Taiwan's National Health Insurance Database. A total of 7556 newly diagnosed patients with PCN and 7556 individuals without PCN were included in the PCN and comparison cohort (PC and CC), respectively, between 2008 and 2018, with propensity score matching for age, gender, comorbidity, medication, and date of PCN diagnosis. The occurrence of AMI was monitored until the end of 2019, and AMI risk was assessed using Cox proportional hazard regression models. Results: The overall incidence of AMI was 1.34-fold higher in the PC than in the CC (4.33 vs. 3.23 per 1000 person-years, respectively, $p < 0.05$), with an adjusted hazard ratio (aHR) of 1.36 (95% confidence interval (CI): 1.08–1.72) after controlling for age, gender, comorbidity, and medication. Further analyses showed a higher risk of AMI with increased annual number of emergency department visits among patients with PCN (aHR: 1.30, 95% CI: 1.01–1.66 (<1) and aHR: 1.68, 95% CI: 1.13–2.50 ($\geq$1)). Conclusion: Patients with PCN had a significantly higher risk of developing AMI than those without PCN. Clinicians should pay more attention to prevent AMI episodes in patients with PCN.

Keywords: interstitial lung disease (ILD); pneumoconiosis; coronary artery disease (CAD); acute myocardial infarction (AMI)

1. Introduction

Pneumoconiosis (PCN) is one of a group of interstitial lung diseases caused by inhalation of certain kinds of dust particles that damage the lungs [1]. PCN can be simple or complicated. Simple PCN causes small and round nodules and complicated PCN causes a lot of scarring and fibrosis tissues in the lungs. Prevalence and incidence of PCN have remained high in recent decades worldwide. There are predicted to be 500,000 alive cases and 60,000 new cases annually. Mortality from PCN also remains high, with more than 20,000 deaths annually [2]. Various kinds of mineral dust could cause various respiratory and cardiovascular diseases. PCN is associated with poor prognosis from these diseases [3–5].

Acute myocardial infarction (AMI) is a medical emergency caused by gross cell death of the cardiac muscle as a result of ischemia, often diagnosed clinically through the cardinal symptoms and signs, electrocardiography (ECG), biochemical testing, non-invasive and invasive imaging techniques, and consecutive pathological evaluation. Approximately 550,000 first episodes and 200,000 recurrent episodes of AMI occur each year in the United States [6]. An atherosclerotic patch dislodges to form coronary plaque, which can compromise the blood vessel patency leading to AMI, which has a high mortality rate. Severe types of AMI are predominantly identified as ST-segment elevation in ECG. Lifestyle, environmental, and genetic factors are contributory factors for AMI. Common risk factors include smoking, diabetes mellitus (DM), hyperlipidemia (HL), and hypertension (HTN). Obesity, homocystinuria, hyperuricemia, and psychosocial stress are also risk factors [7]. However, many potential risk factors are still under investigation.

Several pulmonary and cardiovascular comorbidities have been considered associated with PCN [8–12]. Previously investigated comorbidities included HTN, DM, HL, ischemic heart disease, atrial fibrillation, congestive heart failure, chronic obstructive pulmonary disease (COPD), gastroesophageal reflux disease, and chronic kidney disease (CKD) [3,13]. Most previous studies on the subsequent effects of PCN have focused on interstitial lung disease, pulmonary fibrosis, COPD, and lung cancer. However, blood vessel injuries related to PCN have not been often researched and documented. Using multidetector computed tomography, Lee et al. [14] analyzed the coronary artery calcification (CAC) among 76 patients who were exposed to inorganic dust and reported that PCN was more common among patients with CAC than the non-CAC population. They suggested that additional research is imperative to evaluate the inflammatory reaction of PCN and its influence on atherosclerosis and coronary artery disease (CAD). Since AMI is an important but largely unexplored medical condition, the current study aimed to explore the relationship between PCN and AMI risk.

2. Materials and Methods

2.1. Data Source

Taiwan introduced a single-payer National Health Insurance (NHI) program on 1 March 1995, in which 99.9% of the population has enrolled since 2014. Here, a database comprising the registration files and original claims data of all individuals regarding reimbursement was organized. Many databases resulting from this database are provided to scientists in Taiwan for research activities. Data in the NHI can be utilized for locating patients or healthcare providers, such as medical organizations, institutions, and physicians; however, these data were jumbled before being handed over to researchers. Thus, it is theoretically impossible to query the data alone to identify individuals at any level using this database. Citizens of Taiwan who were eligible to research were alone entitled to apply to this database which was exclusively for research use. Applicants must follow the Computer-processed Personal Data Protection Law and associated regulations of Taiwan's Ministry of Health and Welfare, and an agreement must be signed by the applicant and his/her supervisor upon application submission. All applications are reviewed before data are released. Each year, Taiwan's Ministry of Health and Welfare gathers information from the NHI database and sorts them as data files, including registration files and original claims data for reimbursement. These data files are de-coded by jumbling the identification codes of patients and medical facilities and constructing original files of the NHI database. In this study, we used the NHI database, which included 31,488,321 individuals and detailed medical information from 2008 to 2019 (Figure 1). This study was approved by the Research Ethics Committee of China Medical University Hospital (CMUH110-REC3-133).

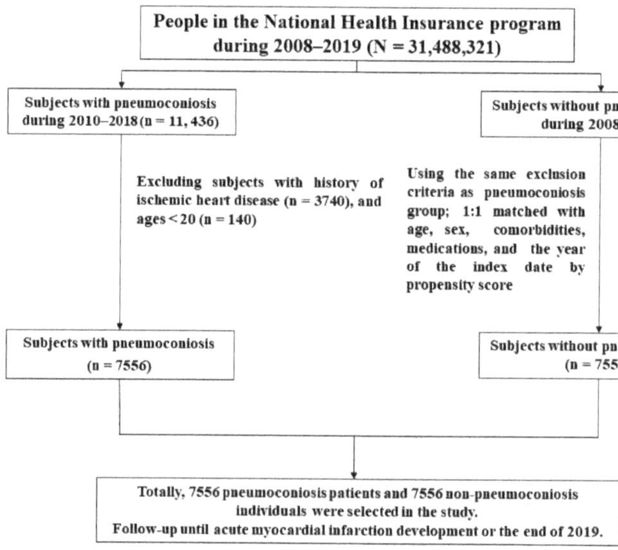

Figure 1. Flowchart of participant selection.

2.2. Study Cohorts

The PCN cohort (PC) enrolled newly diagnosed adult patients between 2008 and 2018 (International Classification of Diseases (ICD) codes 500, 501, 502, 503, 505, J60, J61, J62, J63, and J64). In the present study, the common cause of PCN was inhalation of mineral dust. Mineral dust includes coal dust, asbestos, crystalline silica, and other materials. The detailed codes were coal worker's PCN (ICD code 500 and J60), PCN due to asbestos (ICD code 501 and J61), PCN due to dust containing silica (ICD code 502 and J62), PCN due to other inorganic dusts (ICD code 503 and J63), and unspecified PCN (ICD code 505 and J64). There were some other references using similar definitions [11,15]. We excluded those patients having ischemic heart disease (ICD codes 410–414 and I20–I25) before the diagnosis of PCN. We enrolled participants free of PCN in the comparison cohort (CC), after propensity score matching for age, gender, comorbidity, medication, and date of PCN diagnosis (index date). Individuals who had ischemic heart disease prior to the index date were also excluded. Both cohorts were followed until an AMI episode, withdrawal from insurance system, death, or 31 December 2019.

2.3. Outcome and Related Factors

The primary outcome was AMI (ICD codes 410 and I21), and age, gender, comorbidities, and medication were confounding variables. Hence, various associated comorbidities at baseline, such as HTN (ICD codes 401–405 and I10–I16), DM (ICD codes 250 and E08–E13), HL (ICD codes 272 and E78), asthma/COPD (ICD codes 491, 492, 493, 496, and J41–J45), cerebrovascular disease (CVD, ICD 430–438 and I60–I69), and CKD (ICD 585 and N18) were identified for adjustment. Corticosteroid users were defined as those who used the medication for more than 28 days.

2.4. Statistical Analysis

The proportions of age distribution, gender, comorbidity, and medication in both groups were compared using the Chi-squared test. The mean age was analyzed through Student's *t*-test. The cumulative incidences of AMI in the PC and CC were calculated using the Kaplan–Meier test, and the log-rank test was employed to determine the significance level. Univariable and multivariable Cox proportional hazard regression models were employed to evaluate crude and adjusted hazard ratios (cHRs and aHRs) with 95% confidence intervals (CIs). Data were analyzed using the SAS statistical software (version 9.4 for Windows; SAS Institute, Inc., Cary, NC, USA). A p-value < 0.05 was considered statistically significant.

3. Results

There was a total of 7556 individuals in the PC and 7556 individuals in the CC (Table 1). The proportions of coal worker's PCN, PCN due to dust containing silica, PCN due to asbestos, and PCN due to other or unspecified dust were 38.3%, 4.7%, 1.5%, and 55.5%, respectively. The mean age was 66.9 ± 13.1 years in the PC and 66.8 ± 13.3 years in the CC. Both cohorts consisted of 85.8% males. The extents of comorbidities and medication in the PC and CC were as follows: HTN (43.6% vs. 44.0%), DM (17.9% vs. 18.4%), HL (23.9% vs. 24.3%), asthma/COPD (44.3% vs. 44.9%), CVD (11.56% vs. 9.23%), CKD (4.55% vs. 4.98%), and corticosteroid use (45.3% vs. 45.3%). There was no significant difference in age, gender, comorbidities, and corticosteroid usage between the PC and CC. The average follow-up time was 4.74 ± 2.87 and 5.25 ± 2.73 years in the PC and CC, respectively, during which the increasing incidence of AMI was remarkably greater in the PC than in the CC (Figure 2, p = 0.0133 in the log-rank test).

Table 1. Baseline characteristics in the study population.

	Pneumoconiosis				
	No N = 7556		Yes N = 7556		
	n	%	*n*	%	*p*-Value †
Age					0.89
20–49	738	9.8	725	9.6	
50–64	2302	30.5	2324	30.8	
≥65	4516	59.8	4507	59.7	
Mean ± SD	66.8	±13.3	66.9	±13.1	0.70
Gender					0.96
Women	1070	14.2	1072	14.2	
Men	6486	85.8	6484	85.8	
Comorbidity					
HTN	3327	44.0	3297	43.6	0.62
DM	1392	18.4	1350	17.9	0.38
HL	1837	24.3	1806	23.9	0.56
Asthma/COPD	3391	44.9	3348	44.3	0.48
CVD	1055	14.0	1035	13.7	0.64
CKD	376	4.98	344	4.55	0.22
Medication					1.00
Corticosteroid	3425	45.3	3425	45.3	

CKD = chronic kidney disease; COPD = chronic obstructive pulmonary disease; CVD = cerebrovascular disease; DM = diabetes mellitus; HL = hyperlipidemia; HTN = hypertension; SD = standard deviation. † Chi-squared test and t-test.

The overall occurrence of AMI was 1.34 times greater in the PC than in the CC (4.33 vs. 3.23 per 1000 person-years, respectively, p < 0.05, Table 2), with an aHR of 1.36 (95% CI: 1.08–1.72) following controlling for age, gender, comorbidity, and medication usage. The aHRs of AMI were 2.95 times greater in patients aged between 50–64 years (95% CI: 1.35–6.44) and 4.40 times greater in those aged 65 years and above (95% CI: 2.05–9.45) than in those aged between 20–49 years. The aHR of AMI was 2.84 times greater in males than in females (95% CI: 1.71–4.72). Further, the AMI risk was remarkably higher in participants with HTN (aHR: 1.63, 95% CI: 1.25–2.11) and DM (aHR: 1.57, 95% CI: 1.17–2.10) than in those without. The incidence rates of AMI were also greater among patients with HL, asthma/COPD, CVD, and CKD than those without, but without statistical significance.

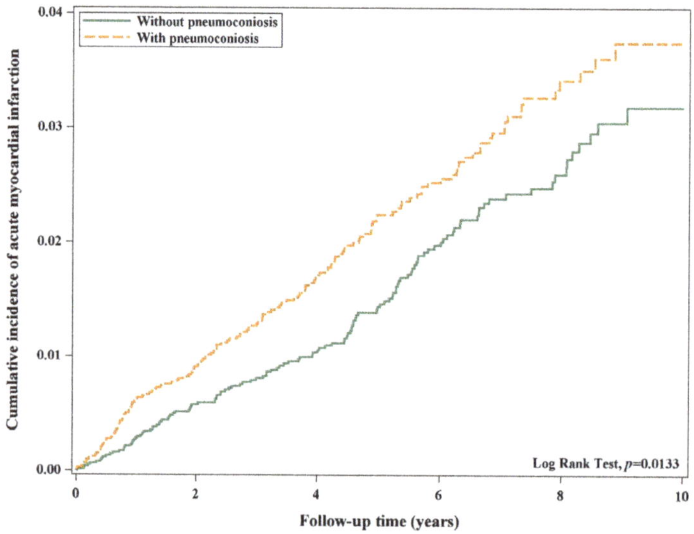

Figure 2. Cumulative incidence of acute myocardial infarction in the pneumoconiosis cohort (PC) and comparison cohort (CC).

Table 2. Risk factors of acute myocardial infarction among the study participants.

	Event	PY	IR †	Crude HR (95% CI)	Adjusted HR # (95% CI)
Pneumoconiosis					
No	128	39,675	3.23	1 (Reference)	1 (Reference)
Yes	155	35,799	4.33	1.34 (1.06–1.70) *	1.36 (1.08–1.72) *
Age					
20–49	7	8418	0.83	1 (Reference)	1 (Reference)
50–64	70	24,990	2.80	3.36 (1.55–7.31) **	2.95 (1.35–6.44) **
≥65	206	42,066	4.90	5.90 (2.78–12.5) ***	4.40 (2.05–9.45) ***
Gender					
Women	16	11,136	1.44	1 (Reference)	1 (Reference)
Men	267	64,339	4.15	2.89 (1.75–4.79) ***	2.84 (1.71–4.72) ***
Comorbidity					
HTN					
No	117	44,428	2.63	1 (Reference)	1 (Reference)
Yes	166	31,046	5.35	2.04 (1.61–2.58) ***	1.63 (1.25–2.11) ***
DM					
No	209	63,228	3.31	1 (Reference)	1 (Reference)
Yes	74	12,246	6.04	1.84 (1.41–2.39) ***	1.57 (1.17–2.10) **
HL					
No	206	57,784	3.57	1 (Reference)	1 (Reference)
Yes	77	17,690	4.35	1.22 (0.94–1.59)	0.93 (0.70–1.25)
Asthma/COPD					
No	148	44,336	3.34	1 (Reference)	1 (Reference)
Yes	135	31,138	4.34	1.30 (1.03–1.65) *	1.22 (0.96–1.55)
CVD					
No	241	67,118	3.59	1 (Reference)	1 (Reference)
Yes	42	8356	5.03	1.41 (1.01–1.95) *	0.94 (0.67–1.32)

Table 2. *Cont.*

	Event	PY	IR †	Crude HR (95% CI)	Adjusted HR # (95% CI)
CKD					
No	263	72,743	3.62	1 (Reference)	1 (Reference)
Yes	20	2731	7.32	2.04 (1.30–3.22) **	1.51 (0.95–2.41)

CI = confidence interval; CKD = chronic kidney disease; COPD = chronic obstructive pulmonary disease; CVD = cerebrovascular disease; DM = diabetes mellitus; HL = hyperlipidemia; HR = hazard ratio; HTN = hypertension; IR = incidence rate; PY = person-years. † Incidence rate per 1000 person-years. # Adjustment for age, gender, comorbidity, and medication; * $p < 0.05$, ** $p < 0.01$, *** $p < 0.001$.

Table 3 shows the incidence and aHR of AMI for the PC and CC by age, gender, and presence of comorbidities. In the <65 and ≥65 years age groups, the age-specific aHRs in the PC compared to the CC were 1.55 (95% CI: 0.98–2.43) and 1.32 (95% CI: 1.00–1.73), respectively. The gender-specific aHRs in the PC compared to the CC were 1.06 (95% CI: 0.40–2.82) in females and 1.38 (95% CI: 1.08–1.76) in males. The comorbidity-specific aHRs in the PC compared to the CC were 1.10 (95% CI: 0.85–1.42) in participants presenting with comorbidity and 3.99 (95% CI: 2.09–7.61) in participants without comorbidity.

Table 3. Incidence rates and hazard ratios of acute myocardial infarction for the pneumoconiosis cohort (PC) compared to the comparison cohort (CC) by age, gender, and comorbidity.

	Pneumoconiosis							
	No			Yes				
	Event	PY	IR †	Event	PY	IR †	Crude HR (95% CI)	Adjusted HR # (95% CI)
Age								
<65	32	17,204	1.86	45	16,203	2.78	1.50 (0.95–2.36)	1.55 (0.98–2.43)
≥65	96	22,470	4.27	110	19,596	5.61	1.32 (1.00–1.73) *	1.32 (1.00–1.73) *
Gender								
Women	8	5673	1.41	8	5463	1.46	1.03 (0.39–2.75)	1.06 (0.40–2.82)
Men	120	34,002	3.53	147	30,336	4.85	1.38 (1.08–1.75) **	1.38 (1.08–1.76) **
Comorbidity ‡								
No	12	10,999	1.09	40	9944	4.02	3.71 (1.95–7.08) ***	3.99 (2.09–7.61) ***
Yes	116	28,675	4.05	115	25,855	4.45	1.10 (0.85–1.42)	1.10 (0.85–1.42)

CI = confidence interval; HR = hazard ratio; IR = incidence rate; PY = person-years. † Incidence rate per 1000 person-years. # Adjustment for age, gender, comorbidity, and medication. ‡ Participants with any comorbidity were selected into the comorbidity group; * $p < 0.05$, ** $p < 0.01$, *** $p < 0.001$.

Table 4 shows that AMI risk was further higher for participants that had increased annual emergency medical demands in the PC compared with CC (aHR: 1.30, 95% CI: 1.01–1.66 (<1) and aHR: 1.68, 95% CI: 1.13–2.50 (≥1)).

Table 4. Events and hazard ratios of acute myocardial infarction associated with mean number of annual emergency department visit for participants with pneumoconiosis compared with those without pneumoconiosis.

	Events	Crude HR (95% CI)	Adjusted HR † (95% CI)
Non-pneumoconiosis	128	1 (reference)	1 (reference)
Pneumoconiosis			
Annual ED visits			
<1	124	1.29 (1.01–1.65) *	1.30 (1.01–1.66) *
≥1	31	1.60 (1.08–2.36) *	1.68 (1.13–2.50) *

CI = confidence interval; ED = emergency department; HR = hazard ratio. † Adjustment for age, sex, comorbidity, and medication; * $p < 0.05$.

4. Discussion

In this population-based, retrospective cohort study that investigated the association between PCN and risk factors for AMI, subjects with PCN had a greater risk of AMI compared to non-PCN subjects. The incidence of AMI was similar between PCN patients with and without any comorbidity. Additionally, the AMI risk was greater for patients with PCN who had proportionally greater frequency of annual emergency department visits. Finally, this cohort highlighted useful epidemiologic data for PCN on a large scale.

The prevalence of CAD among PCN has been presented by Paul et al., who collected information from 8531 patients with PCN from the 5% Medicare Claims Limited Data Set in the United States between 2011 and 2014 [5]. They reported that CAD occurred in 13.9% of all patients with PCN and accounted for 18.7% of those who lost their lives. In their study, Beggs et al. reported patterns of PCN mortality in Kentucky [4]. Of the 330 deaths with PCN, there was a large proportion of heart-related deaths (23.3%), with 37 (12.2%) from cardiac arrest, 16 (4.8%) from myocardial infarction, 16 (4.8%) from congestive heart failure, and 8 (2.4%) from arteriosclerosis. In addition, Hu et al. enrolled 12,209 patients with PCN from the Taiwan NHI database and registered that 2919 (23.9%) of them had CAD [11]. This finding indicates that CAD is a significant comorbidity and cause of death among patients with PCN.

The patho-mechanical connection between PCN and atherosclerosis was illustrated, starting from dust inhalation that triggers a prothrombotic activity via the interleukin-6-dependent system, leading to diminished clotting time, intravascular thrombin formation, and accelerated arterial thrombosis [16]. In addition, exposure to dust particles could increase plasminogen activator inhibitor-1 and suppress tissue plasminogen activator, causing impaired fibrinolysis [17]. Moreover, dust exposure could suppress the tissue factor pathway inhibitor and enhance the extrinsic coagulation pathway to exacerbate intravascular thrombosis [18]. Overall, PCN-related inflammatory reaction, blood vessel injury, and thrombosis-embolic activity might play major roles in the formation of atherosclerosis and following AMI [19,20].

It is crucial to understand the differences in cardiovascular comorbidities in people with and without PCN, which is pivotal in the development of AMI. In the study, we performed propensity score matching for the comorbidities, and thus, no difference was observed for all comorbidities between the PCN group and the comparison group. However, in several studies that did not match for comorbidities, we can find the difference in cardiovascular comorbidities between people with and without PCN. Shen et al. [8] collected data of 3374 patients with PCN from the Registry of Catastrophic Illness database and 13,496 individuals without PCN from the Longitudinal Health Insurance Database in Taiwan. They found that the magnitudes of cardiovascular comorbidities in the PC and CC were as follows: HTN (48.96% vs. 48.47%, $p = 0.62$), DM (13.63% vs. 16.28%, $p < 0.001$), HL (20.45% vs. 21.83%, $p = 0.09$), asthma (30.32% vs. 7.48%, $p < 0.001$), COPD (61.65% vs. 5.42%, $p < 0.001$), heart failure (6.85% vs. 3.81%, $p < 0.001$), CVD (6.37% vs. 5.79%, $p = 0.22$), and CKD (1.72% vs. 1.99%, $p = 0.35$). Cheng et al. [9] established a PC ($n = 1238$) from the Registry of Catastrophic Illness database and a CC ($n = 4952$) from the Longitudinal Health Insurance Database; they found that the magnitudes of cardiovascular comorbidities in the PC and CC were as follows: HTN (25.2% vs. 23.8%, $p = 0.305$), DM (10.1% vs. 11.3%, $p = 0.217$), HL (10.3% vs. 8.4%, $p = 0.039$), COPD (25.0% vs. 11.1%, $p < 0.001$), and CKD (0.6% vs. 1.2%, $p = 0.119$). Chuang et al. [10] obtained data of 6940 patients with PCN from the Registry of Catastrophic Illness database and 27,760 individuals without PCN from the Longitudinal Health Insurance Database; they found that the proportions of cardiovascular comorbidities in the PC and CC were as follows: HTN (34.5% vs. 34.3%, $p = 0.84$), DM (4.97% vs. 6.45%, $p < 0.001$), HL (11.4% vs. 13.7%, $p < 0.001$), COPD (50.6% vs. 8.79%, $p < 0.001$), heart failure (1.76% vs. 0.79%, $p < 0.001$), and atrial fibrillation (0.43% vs. 0.34%, $p = 0.24$). Yen et al. [12] enrolled 8923 patients with PCN from the Registry of Catastrophic Illness database and 35,692 individuals without PCN from the Longitudinal Health Insurance Database; they noted that the proportions of cardiovascular comorbidities in the PC and CC were as follows: HTN (42.3% vs. 44.9%, $p < 0.001$), DM (6.52% vs. 8.79%, $p < 0.001$), HL (15.8% vs. 18.4%, $p < 0.001$), COPD (52.4% vs. 11.7%, $p < 0.001$), and CVD (4.81% vs. 5.32%, $p = 0.05$). In the comorbidity analyses in these studies, we found that HTN, DM, and HL had heterogeneous differences between those with and without PCN when in different study designs; COPD and heart failure had persistent predominantly higher prevalence in patients with PCN than those without PCN. Additionally, the present study used the entire population (n = 31,488,321) in the same NHI database to identify those with and without PCN, and it was believed to have more precise results.

The large PC and absolutely matched cohort used for comparison in this study should be rated as the first strength of this retrospective cohort study. Almost every subject accomplished the follow-up. A prospective cohort study might be expensive, hence, was not feasible; thus, an alternative less expensive retrospective cohort study was preferred utilizing Taiwan's NHI database [21–23]. In addition, this cohort reflected a "real world" scenario in which PCN, AMI, and all comorbidities were clinically and physically assessed during medical consultation.

The operational definitions for PCN, AMI, and comorbidities were adopted from ICD codes, and all diagnoses were dependent on the skill and competence level of clinicians. If this factor had influenced the study results confoundingly, that might be considered a limitation. Detailed information on weight, body mass index, history of smoking, occupational hazard exposure, and family history were lacking in the database. This is also a limitation, as all of them are relevant risk factors for AMI. Third, related clinical variables such as laboratory data (cardiac enzymes, b-type natriuretic peptide, glycated hemoglobin, hemoglobin, etc.), pulmonary function tests, ECG, echocardiography, chest radiography and computed tomography results, and pathology findings were unavailable [15].

5. Conclusions

Patients with PCN had a significantly greater risk for AMI compared to subjects without PCN. The incidence of AMI was similar between PCN patients with and without any comorbidity. Clinicians should pay attention to prevent AMI episodes in patients with PCN, even in those without obvious cardiovascular comorbidity.

Author Contributions: Conceptualization, J.-H.C., T.-C.S., K.-W.C., Y.-R.W. and K.-C.C.; Data curation, C.-L.L. and C.Y.H.; Formal analysis, J.-H.C., T.-C.S., K.-W.C., C.-L.L., Y.-R.W. and K.-C.C.; Supervision, Y.-R.W., K.-C.C. and C.Y.H.; Writing—original draft, all authors. All authors have read and agreed to the published version of the manuscript.

Funding: The study was supported by Taiwan Ministry of Health and Welfare Clinical Trial Center (MOHW111-TDU-B-212-124004) and China Medical University Hospital (DMR-111-105).

Institutional Review Board Statement: The study was conducted in accordance with the Declaration of Helsinki and approved by the Ethics Committee of China Medical University Hospital (CMUH110-REC3-133).

Informed Consent Statement: Informed consent was waived due to the data extracted from the National Health Insurance Database, which provide only comprehensive de-identified healthcare information.

Data Availability Statement: All data generated or analyzed during this study are included in this manuscript.

Acknowledgments: We are grateful to the MOHW Health and Welfare Data Science Center at China Medical University for providing administrative and technical support. We would like to offer special thanks to Chung Y Hsu, who, although no longer with us, contributes to National Health Insurance database research in Taiwan.

Conflicts of Interest: The authors declare no conflict of interest.

References

1. Qi, X.M.; Luo, Y.; Song, M.Y.; Liu, Y.; Shu, T.; Liu, Y.; Pang, J.L.; Wang, J.; Wang, C. Pneumoconiosis: Current status and future prospects. *Chin. Med. J.* **2021**, *134*, 898–907. [CrossRef] [PubMed]
2. GBD 2017 Disease and Injury Incidence and Prevalence Collaborators. Global, regional, and national incidence, prevalence, and years lived with disability for 354 diseases and injuries for 195 countries and territories, 1990–2017: A systematic analysis for the Global Burden of Disease Study 2017. *Lancet* **2018**, *392*, 1789–1858. [CrossRef] [PubMed]
3. Hall, N.B.; Blackley, D.J.; Halldin, C.N.; Laney, A.S. Current review of pneumoconiosis among US coal miners. *Curr. Environ. Health Rep.* **2019**, *6*, 137–147. [CrossRef] [PubMed]
4. Beggs, J.A.; Slavova, S.; Bunn, T.L. Patterns of pneumoconiosis mortality in Kentucky: Analysis of death certificate data. *Am. J. Ind. Med.* **2015**, *58*, 1075–1082. [CrossRef] [PubMed]
5. Paul, R.; Adeyemi, O.; Arif, A.A. Estimating mortality from coal workers' pneumoconiosis among Medicare beneficiaries with pneumoconiosis using binary regressions for spatially sparse data. *Am. J. Ind. Med.* **2022**, *65*, 262–267. [CrossRef]
6. Anderson, J.L.; Morrow, D.A. Acute myocardial infarction. *N. Engl. J. Med.* **2017**, *376*, 2053–2064. [CrossRef]
7. Malakar, A.K.; Choudhury, D.; Halder, B.; Paul, P.; Uddin, A.; Chakraborty, S. A review on coronary artery disease, its risk factors, and therapeutics. *J. Cell. Physiol.* **2019**, *234*, 16812–16823. [CrossRef]
8. Shen, C.H.; Lin, T.Y.; Huang, W.Y.; Chen, H.J.; Kao, C.H. Pneumoconiosis increases the risk of peripheral arterial disease: A nationwide population-based study. *Medicine* **2015**, *94*, e911. [CrossRef]
9. Cheng, Y.Y.; Hsu, K.H.; Chen, Y.H.; Lin, C.H. Increased risk of ischemic stroke in patients with pneumoconiosis. *J. Clin. Neurosci.* **2015**, *22*, 363–367. [CrossRef]
10. Chuang, C.S.; Ho, S.C.; Lin, C.L.; Lin, M.C.; Kao, C.H. Risk of cerebrovascular events in pneumoconiosis patients: A population-based study, 1996–2011. *Medicine* **2016**, *95*, e2944. [CrossRef]
11. Hu, W.S.; Lin, C.L. Risk of atrial fibrillation in patients with pneumoconiosis: A nationwide study in Taiwan. *Clin. Cardiol.* **2020**, *43*, 66–70. [CrossRef] [PubMed]
12. Yen, C.M.; Lin, C.L.; Lin, M.C.; Chen, H.Y.; Lu, N.H.; Kao, C.H. Pneumoconiosis increases the risk of congestive heart failure: A nationwide population-based cohort study. *Medicine* **2016**, *95*, e3972. [CrossRef] [PubMed]
13. Sun, B.Q.; Zhao, H.L.; Xie, Y. Progress in epidemiological studies on pneumoconiosis with comorbidities. *Zhonghua Lao Dong Wei Sheng Zhi Ye Bing Za Zhi* **2021**, *39*, 389–393.
14. Lee, W.J.; Shin, J.H.; Park, S.Y. Relation of pulmonary function impairment and coronary artery calcification by multi-detector computed tomography in group exposed to inorganic dusts. *Tuberc. Respir. Dis.* **2013**, *74*, 56–62. [CrossRef]
15. Wei, C.H.; Li, C.H.; Shen, T.C.; Hung, Y.T.; Tu, C.Y.; Hsia, T.C.; Hsu, W.-H.; Hsu, C.Y. Risk of chronic kidney disease in pneumoconiosis: Results from a retrospective cohort study (2008–2019). *Biomedicines* **2023**, *11*, 150. [CrossRef]
16. Mutlu, G.M.; Green, D.; Bellmeyer, A.; Baker, C.M.; Burgess, Z.; Rajamannan, N.; Christman, J.W.; Foiles, N.; Kamp, D.W.; Ghio, A.J.; et al. Ambient particulate matter accelerates coagulation via an IL-6-dependent pathway. *J. Clin. Investig.* **2007**, *117*, 2952–2961. [CrossRef]
17. Budinger, G.R.; McKell, J.L.; Urich, D.; Foiles, N.; Weiss, I.; Chiarella, S.E.; Gonzalez, A.; Soberanes, S.; Ghio, A.J.; Nigdelioglu, R.; et al. Particulate matter-induced lung inflammation increases systemic levels of PAI-1 and activates coagulation through distinct mechanisms. *PLoS ONE* **2011**, *6*, e18525. [CrossRef]
18. Wu, Z.; Liu, M.C.; Liang, M.; Fu, J. Sirt1 protects against thrombomodulin down-regulation and lung coagulation after particulate matter exposure. *Blood* **2012**, *119*, 2422–2429. [CrossRef]
19. Zheng, Y.; Liang, L.; Qin, T.; Yang, G.; An, S.; Wang, Y.; Li, Z.; Shao, Z.; Zhu, X.; Yao, T.; et al. Cross-section analysis of coal workers' pneumoconiosis and higher brachial-ankle pulse wave velocity within Kailuan study. *BMC Public Health* **2017**, *17*, 148. [CrossRef]
20. Gellissen, J.; Pattloch, D.; Möhner, M. Effects of occupational exposure to respirable quartz dust on acute myocardial infarction. *Occup. Environ. Med.* **2019**, *76*, 370–375. [CrossRef] [PubMed]
21. Hsing, A.W.; Ioannidis, J.P. Nationwide population science: Lessons from the Taiwan National Health Insurance Research Database. *JAMA Intern. Med.* **2015**, *175*, 1527–1529. [CrossRef] [PubMed]
22. Lai, S.W.; Lin, C.L.; Liao, K.F. Risk of contracting pneumonia among patients with pre-dialysis chronic kidney disease: A population-based cohort study in Taiwan. *Biomedicine* **2017**, *7*, 20. [CrossRef] [PubMed]
23. Hsieh, C.Y.; Su, C.C.; Shao, S.C.; Sung, S.F.; Lin, S.J.; Kao Yang, Y.H.; Lai, E.C.-C. Taiwan's National Health Insurance Research Database: Past and future. *Clin. Epidemiol.* **2019**, *11*, 349–358. [CrossRef] [PubMed]

biomedicines

Review

$K_{Ca}2$ and $K_{Ca}3.1$ Channels in the Airways: A New Therapeutic Target

Razan Orfali [1,2], Ali AlFaiz [2], Mohammad Asikur Rahman [1], Liz Lau [1], Young-Woo Nam [1,*] and Miao Zhang [1,*]

1 Department of Biomedical and Pharmaceutical Sciences, Chapman University School of Pharmacy, Irvine, CA 92618, USA; rorfali@kfmc.med.sa (R.O.)
2 Biomedical Research Administration, Research Centre, King Fahad Medical City, Riyadh Second Health Cluster, Riyadh 12231, Saudi Arabia
* Correspondence: ywnam@chapman.edu (Y.-W.N.); zhang@chapman.edu (M.Z.)

Abstract: K^+ channels are involved in many critical functions in lung physiology. Recently, the family of Ca^{2+}-activated K^+ channels (K_{Ca}) has received more attention, and a massive amount of effort has been devoted to developing selective medications targeting these channels. Within the family of K_{Ca} channels, three small-conductance Ca^{2+}-activated K^+ ($K_{Ca}2$) channel subtypes, together with the intermediate-conductance $K_{Ca}3.1$ channel, are voltage-independent K^+ channels, and they mediate Ca^{2+}-induced membrane hyperpolarization. Many $K_{Ca}2$ channel members are involved in crucial roles in physiological and pathological systems throughout the body. In this article, different subtypes of $K_{Ca}2$ and $K_{Ca}3.1$ channels and their functions in respiratory diseases are discussed. Additionally, the pharmacology of the $K_{Ca}2$ and $K_{Ca}3.1$ channels and the link between these channels and respiratory ciliary regulations will be explained in more detail. In the future, specific modulators for small or intermediate Ca^{2+}-activated K^+ channels may offer a unique therapeutic opportunity to treat muco-obstructive lung diseases.

Keywords: $K_{Ca}2$ channels; lungs; motile cilia; cystic fibrosis; anosmia; chronic obstructive pulmonary diseases

1. Introduction

The epithelial surface of the respiratory tract between the nose and the alveoli is constantly exposed to potentially harmful pathogens, particulates, and gaseous materials [1–3]. In response to these challenges, the human body utilizes a series of defense mechanisms to protect the airways, and the primary defense mechanism in the lung is mucociliary clearance (MCC) [1,4]. MCC is a process of specialized organelles called cilia that beat in metachronal waves to impel pathogens and particles trapped by the mucous layer out of the airways. Cilia within the mucociliary system present critical functions in human health; abnormalities in each compartment of the mucociliary system could compromise the mucus clearance process and lead to chronic lung disease [2,3]. Mucociliary dysfunction is commonly associated with chronic airway diseases, and it is one of the pathological observations in patients with cystic fibrosis, primary ciliary dyskinesia, chronic bronchitis, and asthma [5,6]. Airway diseases with associated mucociliary dysfunction remain largely unaddressed, despite the therapeutic progress in treating inflammatory lung diseases [5].

The lung's lining is covered by a thin layer of fluid called airway surface liquid (ASL); it separates the airway epithelium's luminal surface from the external environment. ASL is mainly composed of water, electrolytes, and mucins; it is essential for normal airway function, particularly for proper MCC [3,7,8]. ASL epithelia contain various cell types with distinct morphologies and functions. Of the cell population in the trachea, approximately 60% are ciliated cells; these cells also retain other important roles other than coordinating ciliated movements, such as regulating ion transfer [1,9].

There are detections of over 30 diverse K^+ channels in the airway epithelia, and these K^+ channels maintain the electrochemical gradient and support lung ion and fluid homeostases [1,10–12]. A large portion of airway chloride secretion occurs through the apically located bicarbonate and chloride channels [10]. K^+ channels are involved in many vital functions in lung physiology, such as oxygen sensing, inflammatory responses, enhancing Cl^- transport, and respiratory epithelia repair [9,13]. The basolateral K^+ channel has known regulation effects on Na^+ absorption; reduced Na^+ absorption in the lung shows improvement in muco-obstructive disease. A large portion of airway chloride secretion occurs through the apically located bicarbonate and chloride channels, significantly influenced by some Ca^{2+}-activated K^+ channels (K_{Ca}) that are located apically in the lung [10]. Hence, the specific K^+ group K_{Ca} also regulates MCC and ASL volumes [1]. The small-conductance $K_{Ca}2$ channels and intermediate-conductance $K_{Ca}3.1$ channels are voltage-independent and activated solely by the elevation of the intracellular Ca^{2+} concentration. In this context, we will discuss current knowledge of the functional roles of $K_{Ca}2$ and $K_{Ca}3.1$ channels in the respiratory tract, focusing on their physiological roles in respiratory diseases.

2. Introduction to K_{Ca} Channels

There are several kinds of K^+ channels present in the respiratory epithelium lining airways, and the most indispensable K^+ channels in airway epithelial cells are the Ca^{2+}-activated K^+ channels. They serve as the cell crossroad where Ca^{2+} influx, other ion outfluxes, and membrane potential, all processes governed by K_{Ca} channels, integrate to modulate an extensive array of cellular processes [14]. K_{Ca} channels are subdivided into three major groups, according to their single-channel conductance: large conductance (150–300 pS) K^+ channels (BK or $K_{Ca}1.1$), small conductance (2–20 pS) K^+ channels (SK or $K_{Ca}2$), and intermediate conductance (20–60 pS) K^+ channels (IK or $K_{Ca}3.1$) [15–17]. Each group has specific distinct biophysical and pharmacological properties [18]. $K_{Ca}2.x$ and $K_{Ca}3.1$ channels are voltage-independent and activated exclusively by intracellular Ca^{2+} via the calmodulin (CaM) that is typically bound to these channels and serves as their Ca^{2+} sensor [19]. $K_{Ca}2x$ and $K_{Ca}3.1$ channels, before their cloning, were referred to as small-conductance (SK) or intermediate-conductance (IK) Ca^{2+}-activated K^+ channels, based on their singular conductance of ~10 pS or ∼40 pS in symmetrical solutions to differentiate them from the large-conductance potassium (BK) channel [19,20].

Four *mammalian KCNN* channel subtypes are encoded by the *KCNN* genes, including *KCNN1* for $K_{Ca}2.1$, *KCNN2* for $K_{Ca}2.2$, *KCNN3* for $K_{Ca}2.3$ [21], and *KCNN4* for $K_{Ca}3.1$ [22], respectively [23] (Table 1).

Table 1. Apparent Ca^{2+} sensitivity, structural studies, amino acid sequences alignments and identities between $K_{Ca}2/3$ channel subtypes.

<table>
<tr><th>$K_{Ca}2/3$ α Subunit</th><th>Amino Acids</th><th>Apparent Ca^{2+} Sensitivity (μM)</th><th>$K_{Ca}2$ Subtypes Structural Studies</th><th>Sequence Alignment among $K_{Ca}2$ and $K_{Ca}3.1$ Channels</th></tr>
<tr><td>$K_{Ca}2.1$</td><td>543 [24]</td><td>~0.31 [25,26]</td><td>[27,28]</td><td>$K_{Ca}2.1$ and $K_{Ca}3.1$ share a 43.3% sequence identity [29]</td></tr>
<tr><td>$K_{Ca}2.2$</td><td>579 [25]</td><td>~0.32 [25,30]</td><td>[15,26,29,31]</td><td>$K_{Ca}2.2$ and $K_{Ca}3.1$ share a 45% sequence identity [32]</td></tr>
<tr><td>$K_{Ca}2.3$</td><td>731 [25]</td><td>~0.30 [33,34]</td><td>[34,35]</td><td rowspan="2">$K_{Ca}2.3$ and $K_{Ca}3.$ share a 46.6% sequence identity [34]</td></tr>
<tr><td>$K_{Ca}3.1$</td><td>427 [25]</td><td>~0.27 [33,34]</td><td>[36]</td></tr>
</table>

$K_{Ca}2$ and $K_{Ca}3.1$ Channel Structures

$K_{Ca}2$ and $K_{Ca}3.1$ channels are assembled as homotetramers of four α-subunits; each subunit is composed of six transmembrane α-helical domains denoted as S1–S6 (Figure 1). The selectivity filter within the channel pore between the S5 and S6 transmembrane domains is responsible for the selective permeability of the K^+ ions [30,34]. The $K_{Ca}2/K_{Ca}3.1$ channel subtypes are highly homologous in their six transmembrane domains, but the amino acid sequences and lengths at their cytoplasmic N- and C-termini differ among the subtypes (Table 1) [37].

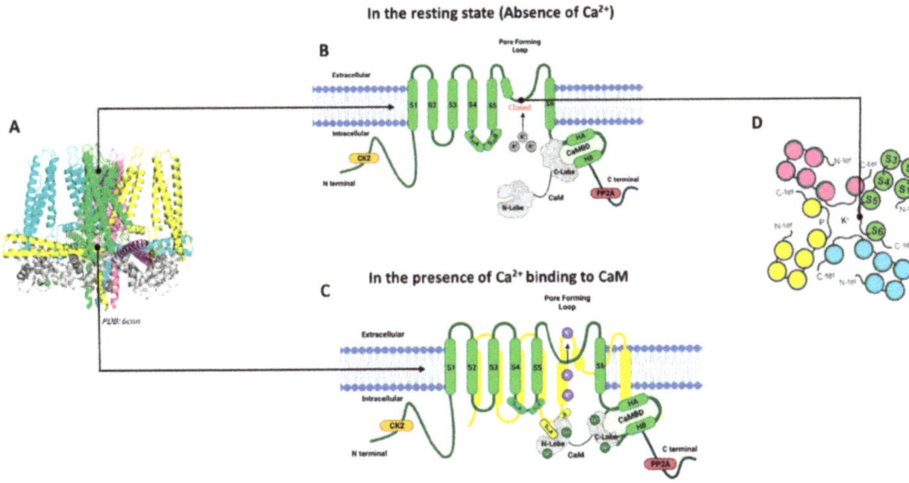

Figure 1. K_{Ca}3.1 and K_{Ca}2 Channel Structures in the presence and absence of Ca^{2+}. K_{Ca}2 and K_{Ca} 3.1 channels are assembled as homotetramers of four α-subunits. (**A**) Human K_{Ca}3.1 channel cryo-EM structure (PDB: 6cnn). For clarity, four-channel subunits are shown in different colors: green, blue, yellow, and purple, along with calmodulin (CaM) (gray). (**B**) Schematic representation of one channel subunit in the absence of Ca^{2+}. (**C**) Schematic representation of one channel subunit in the presence of Ca^{2+}. (**D**) Extracellular top view of the K_{Ca}3.1 and K_{Ca}2 channels. (**A**) was generated using Biorender.com. (**B**,**C**) were generated using Pymol (Schrödinger, LLC, New York, NY, USA).

Among the four $K_{Ca}2/K_{Ca}3.1$ channel subtypes, the full-length cryogenic electron microscopy (cryo-EM) structure is only available for the K_{Ca}3.1 channel determined in the absence and presence of Ca^{2+}, providing insight into the Ca^{2+}/CaM gating mechanism for these channels [36]. The calmodulin-binding domain consists of two α-helices, HA and HB, whereas the S4–S5 linker includes two α-helices, S_{45}A and S_{45}B. The HA and HB helices from one channel subunit, the S4–S5 linker from a neighboring channel subunit, and calmodulin closely interact with each other (Figure 1). When Ca^{2+} is absent, the C-lobe of CaM binds to the HA/HB helices in the proximal channel C-terminus, the N-lobe of CaM is highly flexible, and the channel pore is closed (Figure 1B). In the presence of Ca^{2+}, the N-lobe of CaM becomes well-structured and interacts with the linker between the S4 and S5 transmembrane domains (S4–S5 linker) of a neighboring α-subunit. The interaction between the Ca^{2+}-bound CaM N-lobe and the S4–S5 linker causes the movement of the S6 transmembrane domain and the opening of the channel pore (Figure 1C) [33,38].

K_{Ca}2 channels are activated by Ca^{2+}, with EC_{50} values ranging from 300 to 750 nM, whereas K_{Ca}3.1 channels exhibit apparent Ca^{2+} sensitivities of 100–400 nM [29,34,39]. K_{Ca}2 and K_{Ca}3.1 channels, therefore, play a critical role in the physiologies of various tissues and disease states [22,40]. The advances in understanding the K_{Ca}3.1 structure [36] (the cryo-electron microscopy of the human homotetrameric *KCNN4* channel) and the resulting improvements in other K_{Ca}2 subtypes modeling [29,30,34] have yet to be used, not only for drug discovery but also for understanding the pathophysiological diseases.

3. K_{Ca} Channels in the Respiratory System

The involvement of K^+ channels has been proposed in respiratory conditions such as asthma, chronic obstructive pulmonary diseases (COPD), and cystic fibrosis (CF) [1,12]. In airway epithelial cells, both Cl^- and K^+ transports rely, to some extent, on Ca^{2+}-dependent channel activity (e.g., K_{Ca} channels) [1]. K_{Ca} channels are important in regulating Cl^-

secretion, MCC, and ASL volumes. $K_{Ca}3.1$ and $K_{Ca}2$ channel subtypes located in the airway epithelia, such as $K_{Ca}2.1$ [41] and $K_{Ca}2.3$ [42], maintain the electrochemical gradient and thus support lung ion and fluid homeostasis [1]. Table 2 summarizes the *KCNN* gene family, tissue distribution, physiological roles, and their roles in the lungs.

Table 2. The *KCNN* gene family. Human chromosomal location, tissue distribution, functional effects, and their roles in the lungs.

$K_{Ca}2/3$ α Subunit	Gene	Other Names	Human Chromosomal Location	Tissue Distribution	Physiological Roles	Role in the Lungs
$K_{Ca}2.1$	*KCNN1*	SK1	19p13.11 [25]	Brain [25] Heart [43] Lung [41]	The $K_{Ca}2$ channels underlie the medium AHP and regulate neuronal firing frequency [23,44].	ND *
$K_{Ca}2.2$	*KCNN2*	SK2	5q22.3 [25]	Brain and heart Adrenal gland, lungs, prostate, bladder, and liver [25,45].		ND *
$K_{Ca}2.3$	*KCNN3*	SK3	1q21.3 [25]	Brain and heart Vascular endothelium, lungs, and bladder [25,44]	$K_{Ca}2.3$ and $K_{Ca}3.1$ mediate the endothelium-derived hyperpolarization response [33,46]	(+) $K_{Ca}2.3$ relaxes the pulmonary arteries and bronchi ** [32]
$K_{Ca}3.1$	*KCNN4*	SK4 IK	19q13.31 [25]	Vascular endothelium T and B lymphocytes Microglia, placenta, colon, and red blood cells Lungs and bladder [25,44]	$K_{Ca}3.1$ channels regulate calcium signaling, cellular activation, and cell volume [23,44]	(−) $K_{Ca}3.1$ reduces Na^+ absorption ***, (+) CBF, and MCC [5]. (+) $K_{Ca}3.1$ relaxes the pulmonary arteries and bronchi [42]

* ND: not determined specifically in the respiratory system. ** (+): Activation. *** (−): Inhibition.

3.1. K_{Ca} Channels and the Respiratory Cilia

The $K_{Ca}2$ and $K_{Ca}3.1$ channels are tetramers, and each subunit comprises six transmembrane alpha-helical domains (six TMD), indicated as S1–S6 in each channel subunit. The selectivity of potassium ions across these channels is based on the pore-forming P-loop between the transmembrane S5 and S6 domains. $K_{Ca}2/K_{Ca}3.1$ are more sensitive to Ca^{2+} due to calmodulin CaM acting as a Ca^{2+} sensor (Figure 1) [26,47]. CaM is present in all eukaryotic cells, facilitating various cellular signaling processes, such as the modulation of ion channel actions, regulation of enzymatic activities, and gene expression [14,48]. The ciliary beat of the airway epithelium is believed to be regulated by the level of intracellular Ca^{2+} [49]. The association with calmodulin in the regulation of ciliary beats has been reported as the most important intraciliary Ca^{2+} binding protein [49,50]. Moreover, the activation of $K_{Ca}2$ channels in non-excitable cells, such as epithelial cells, increases Ca^{2+} entry through non-voltage-gated Ca^{2+} channels, thereby increasing intracellular Ca^{2+} concentration [51]. This elevation of intracellular Ca^{2+} is one of the primary regulators of ciliary movement [52]. Thus, $K_{Ca}2$ and $K_{Ca}3.1$ channels will regulate respiratory ciliary activities as part of a complex signaling network.

3.1.1. K_{Ca} Channels and Ciliary Beat Frequency

In vitro measurements of the changes in the CBF of human respiratory cells indicate that Ca^{2+} ionophore speeds the CBF of human respiratory cells mediated through a calmodulin-sensitive system [53]. Airway epithelial cells contain 100 nM of free Ca^{2+} in their cytoplasm, but ciliated cells bear a higher concentration at baseline than club cells [54]. This supports the idea that $K_{Ca}2$ channels may be active during normal conditions in specific airway cells, as these channels show a high sensitivity to Ca^{2+} (Table 2). Significantly, in CF mouse airways, a previous study by Vega et al. [5] determined that *KCNN4*-silencing enhanced MCC when Na^+ absorption was decreased. Additionally, CBF was also increased by $K_{Ca}3.1$ inhibition. An explanation is that $K_{Ca}3.1$ inhibition reduces Na^+ absorption in CF, thereby increasing CBF speeds by hyperpolarizing the apical membrane [5,55].

3.1.2. $K_{Ca}2$ Channels and Cilium Length

Muco-obstructive lung disease is considered the primary cause of morbidity and is responsible for 80% of mortality [55]. The presence of $K_{Ca}2$ channels in a human bron-chial epithelial cell, and structural similarities in the groups of $K_{Ca}2$ and $K_{Ca}3.1$, pro-vides a new direction in the investigating the expression and function of $K_{Ca}2$ channel subtypes in the ciliated human lung epithelial cells. Optimal MCC requires mucus, cilia, and a thin layer of ASL to facilitate ciliary beating [2]. Maintaining a normal range of respiratory cilia length (4 to 7 μm, depending on the airway region) is critical for adequate mucociliary clearance [56]. A qualitative difference exists between short and longer cilia waveform shapes [57], and various acquired lung disorders are marked by abnormalities in both cilia structure and function [56]. Our previous work determined the critical role of $K_{Ca}2.3$ channels in regulating the primary cilia in endothelial cells [58]. Taking advantage of the previous results could help to connect $K_{Ca}2$ channels and respiratory cilia, two crucial components in the Ca^{2+} signaling network of airway epithelial and smooth muscle cells, with potential implications in the pathogenesis of airway diseases.

4. Expression and Physiological Functions of $K_{Ca}2$ and $K_{Ca}3.1$ Channels in the Airways

Many human cells express K_{Ca} channels that have the exceptional ability to trans-late changes in the level of the intracellular second messenger, Ca^{2+}, to changes in membrane K^+ conductance and, thus, resting potential membrane. While K_{Ca} channel subtypes are all regulated by intracellular Ca^{2+}, they are otherwise quite distinct entities, differing in tissue distribution and functions [59]. $K_{Ca}2$ channel subtypes, for example, are widely expressed in the nervous system, where they are involved in regulating the firing frequency of various neurons. On the other hand, the $K_{Ca}3.1$ channel subtype is expressed in peripheral cells, including the erythrocytes and lymphocytes, and has been determined in numerous cancer cells where they have been implicated in growth control [60,61]. Here we demonstrate the expressions and physiological roles of $K_{Ca}2$ and $K_{Ca}3.1$ channels in the airways.

4.1. Expression and Functions of $K_{Ca}2$ in the Respiratory Epithelia

$K_{Ca}2$ channels are widely expressed in various tissues and play an important role in modulating excitable and non-excitable cells. The presence of K_{Ca} channel groups was confirmed at the apical and basolateral membranes of airway epithelial cells [1,62] (Figure 2A). The bronchial epithelium expresses $K_{Ca}2.1$ and $K_{Ca}2.3$ channel subtypes[35,41]. $K_{Ca}2.2$ and $K_{Ca}2.3$ mRNA were detected in the lungs and trachea [3,4]. $K_{Ca}2.2$ and $K_{Ca}2.3$ mRNA were detected in lungs and trachea [6]. $K_{Ca}2.3$ is the only subtype expressed in the pulmonary artery [5]. Figure 2-B shows the major expression sites of $K_{Ca}2$ and $K_{Ca}3.1$ channel subtypes in the airway.

Different ion channels seem to be present in motile cilia [63]. In the nasal cavity, olfactory receptor neurons (ORNs) are adapted to grow various long cilia; they are not motile but can move with the liquid stream of the nasal mucosa to sample odorants entering the nose. The presence of K_{Ca} channel groups in the cilia of ORNs was reported [64].

The involvement of K^+ channels has been proposed in respiratory conditions such as asthma, chronic obstructive pulmonary diseases (COPD), and cystic fibrosis (CF) [1,12]. In airway epithelial cells, both Cl^- and K^+ transports rely, to some extent, on Ca^{2+}-dependent channel activity (e.g., K_{Ca} channels) [1]. K_{Ca} channels are important in regulating Cl^- secretion, MCC, and ASL volumes. $K_{Ca}2$ channel subtypes located in the airway epithelia, such as $K_{Ca}2.1$ [41] and $K_{Ca}2.3$ [42], maintain the electrochemical gradient and thus support lung ion and fluid homeostases [1].

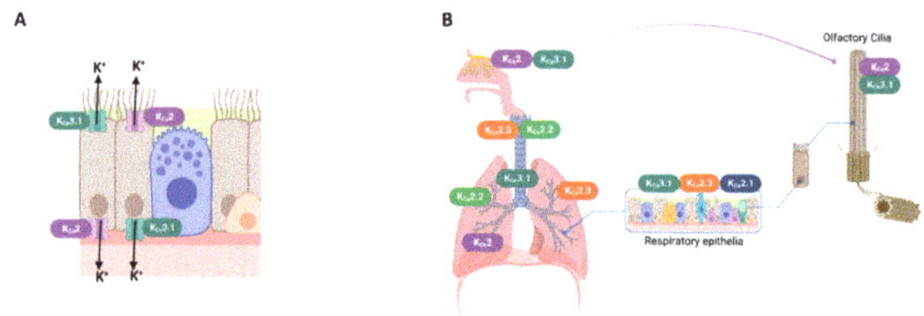

Figure 2. Expression sites of $K_{Ca}2$ and $K_{Ca}3.1$ channels in the respiratory system. (**A**) Schematic drawing of ciliated airway epithelial cells of $K_{Ca}2$ and $K_{Ca}3.1$ channels. (**B**) $K_{Ca}2$ and $K_{Ca}3.1$ channels were expressed in airway smooth muscle [65], airway olfactory nerves [66] and olfactory cilia [64]; $K_{Ca}2.2$ and $K_{Ca}2.3$ subtypes presented in the lungs and the trachea [45]; and $K_{Ca}3.1$, $K_{Ca}2.1$, and $K_{Ca}2.3$ subtypes presented in the respiratory epithelia [41,42]. The pulmonary artery expressed the $K_{Ca}2.3$ subtype [42].

In CF, the equilibrium between Na^+ absorption and Cl^- secretion throughout the airway epithelia is necessary to maintain adequate ASL volume and MCC. The Cl^- secretion in the lungs involves several steps, starting from Cl^- entry through a basolateral channel cotransporter, followed by its exit via apical Cl^- channels, such as the cystic fibrosis transmembrane conductance regulator (CFTR) [3]. The dysfunction of CFTR channels in CF results in decreased Cl^- and fluid secretions and increased Na^+ absorption, leading to inefficient mucociliary clearance and mucus accumulation [67].

In COPD, K_{Ca} channel groups can also act as oxygen sensors for lung diseases, such as COPD associated with pulmonary hypertension [12]. In COPD, pulmonary hypertension is generally believed to be due to hypoxic pulmonary vasoconstriction [68]. K_{Ca} channels potentiated by the low partial pressure of oxygen (PO_2) have been investigated in cerebral resistance myocytes [69]. When hypoxia occurs, K_{Ca} channels activate (preventing repolarization) and relax the pulmonary arteries [70,71].

K_{Ca} channels were proposed as new targets for bronchodilator therapy for chronic diseases such as asthma and COPD [72]. The mentioned COPD-related studies [69–72] examined K_{Ca} channel groups in general. Though one study suggested that human pulmonary artery and bronchial relaxations might be mediated by pharmacological activation of the $K_{Ca}2.3$ channel subtypes [42] (Table 2).

In anosmia, Odorant-induced K^+ conductance is activated by Ca^{2+} [73], and the elevation of intracellular Ca^{2+} is often associated with odorant stimulation in some vertebrates and human olfactory neurons [51,74]. Olfactory Receptor Neurons (ORNs) are located in the nasal epithelia and exhibit spontaneous action potential firing. All K_{Ca} channel groups have been detected in olfactory cilia [52], and the electrophysiological of the whole-cell results confirmed that K_{Ca} channels participate in inhibitory chemo-transduction in the cilia [75]. According to these findings, an apical Ca^{2+} influx opens the K_{Ca} channels, causing membrane hyperpolarization in response to Ca^{2+} influx and thus triggering the inhibition [74].

Moreover, a Ca^{2+} channel blocker, nifedipine, was tested on odorants that induce an inhibitory current in olfactory neurons [74]. This drug effectively abolished the outward current and stimulated the cells with an odorant solution free of nifedipine, and the response was restored [74]. $K_{Ca}2$ channel subtypes which are completely Ca^{2+}-dependent and voltage-independent may play a critical role in treating certain diseases, given the drugs that could target specific ion channels. For example, anosmia could be treated by targeting the $K_{Ca}2$ channel subtypes in olfactory cilia and testing their allosteric modulators [52]. However, the pharmacology of these channels in olfactory neurons has not been fully characterized.

4.2. Expression and Functions of $K_{Ca}3.1$ in the Respiratory Epithelia

The expression of Ca^{2+}-activated potassium (K_{Ca}) channels often correlate positively with cell proliferation. As an example, the expression of $K_{Ca}3.1$ increases 4-fold upon T-lymphocyte activation, and this channel is inhibited with the specific inhibitor that inhibits T-lymphocyte proliferation [76]. This is because the $K_{Ca}3.1$ channel contributes to electrochemical gradients for Ca^{2+} influx, which is critical for the proliferation of the T cells [77]. $K_{Ca}3.1$ is also broadly expressed in other cells of the immune system, such as B cells, macrophages, microglia, and mast cells [78]. The major function of $K_{Ca}3.1$ in immune cells is to hyperpolarize the cell membrane and create the driving force for calcium entry, which is necessary for proliferation, activation, and cytokine production [79]. Previous findings [80] suggest that antigen sensitization up-regulates $K_{Ca}3.1$ expression, which may contribute to enhancing cell migration in response to lymphatic chemokines, particularly in the immunogenic lung dendritic cells subset. Therefore, targeting $K_{Ca}3.1$ crucial for controlling allergic airway inflammation [81] (Table 2 and Figure 2).

K_{Ca} channels have been found to be involved in regulating smooth muscle responses to both contractile and relaxant agonists that elevate intracellular Ca^{2+} [82]. Phenotypic modulation of smooth muscle cells is accompanied by changes in $K_{Ca}3.1$ channel expression characterizing "proliferative" cells [83]. $K_{Ca}3.1$ channels regulate the proliferative responses of vascular smooth muscle cells, fibroblasts, endothelial cells, and T lymphocytes, as well as a some transformed cell types [61,84]. $K_{Ca}3.1$ function is increased by protein kinase A (PKA) [85] and nucleoside diphoshate kinase B (NDPK-B) and inhibited by the histidine phosphatase PHPT1 [86,87]. Since NDPK-B and PHPT1 directly phosphorylate or dephosphorylate $K_{Ca}3.1$ on histidine in the C-terminus, $K_{Ca}3.1$ modulation in mammals is one of the rare examples of histidine kinase/phosphatase regulating a biological process [86].

In allergic lung diseases, $K_{Ca}3.1$ channels regulate Ca^{2+} entry into cells and thereby modulate Ca^{2+}-signaling processes. The entry of positively charged Ca^{2+} into the cells depolarizes the membrane, which limits its own ability to enter the cell through some types of Ca^{2+} channels that are closed at more positive membrane potentials. $K_{Ca}3.1$ activation by elevated intracellular Ca^{2+} maintains a negative membrane potential, which helps to sustain Ca^{2+} entry into the cell. $K_{Ca}3.1$-mediated elevation of intracellular Ca^{2+} is necessary for the production of inflammatory chemokines and cytokines by T cells, mast cells, and macrophages [79,88]. Indeed, proliferation is accompanied by the transcriptional up-regulation of functional $K_{Ca}3.1$ expression and can be inhibited by $K_{Ca}3.1$ inhibitors [86]. It has been reported that the use of $K_{Ca}3.1$ blockers can provide a potential therapeutic target for mast cell-mediated diseases such as asthma [88]. Moreover, blocking $K_{Ca}3.1$ may offer a novel approach to treating idiopathic pulmonary fibrosis [89].

In muco-obstructive hyper tension, the inhibition of the $K_{Ca}3.1$ channel [5] and *Kcnn4* silencing in ion transport and MCC in an animal model of CF/COPD-like muco-obstructive lung disease determined that *Kcnn4* silencing enhances airway disease [5]. The effectiveness of the mucociliary clearance depends mainly on hydration. Water availability in the airways is controlled by transepithelial ion transport. Apical Cl^- secretion and Na^+ absorption play major roles in ASL volume homeostasis [90]. The decline in Na^+ absorption is of potential benefit in muco-obstructive disorders, such as cystic fibroses. It was described earlier in the case of the kidney and intestine, where the inhibition of basolateral K^+ channels decreased Na^+ absorption [5,57], thus supporting the role of K^+ channels on epithelial Na^+ homeostasis.

In pulmonary artery hypertension, elevated pulmonary artery pressure occurs in several diseases, such as asthma, end-stage chronic obstructive pulmonary disease (COPD), and lung fibrosis [66,91,92]. In order to diagnose pulmonary artery hypertension, hemodynamic measurements are taken via right heart catheterization or echocardiography; the condition is defined as a mean pulmonary artery pressure above 25 mmHg at rest or greater than 30 mmHg during normal physical activity [92]. Studies suggest that pharmacologically activating $K_{Ca}3.1$ channels mediates human pulmonary artery and bronchial relaxations [42]

5. Pharmacological $K_{Ca}2$ and $K_{Ca}3.1$ Channel Modulators in Respiratory Diseases

The $K_{Ca}2.3$ and $K_{Ca}3.1$ potassium channels are characterized by their voltage independence, and thus, they are activated by intracellular Ca^{2+}. Due to the distinct distribution of the channel subtypes in the mammalian cells and their involvement in the generation of afterhyperpolarization currents, there has been considerable interest in developing subtype-selective pharmacological tools to study these channels [93,94]. Additionally, $K_{Ca}2.3$ and $K_{Ca}3.1$ channels comprise attractive new targets for several diseases that currently have no effective therapies. The pharmacology of K_{Ca} channels developed relatively rapidly after the cloning of the $K_{Ca}2$ and $K_{Ca}3.1$ channels, as the field now has a wide range of peptides, small-molecule inhibitors, and positive- and negative-gating modulators with differential subtype selectivity available [44].

The $K_{Ca}3.1$ and $K_{Ca}2$ channels have relatively well-developed pharmacological tools. The field now has a wide range of peptides, small-molecule inhibitors, and positive- and negative-gating modulators with differential subtype selectivity available [93]. Table 3 shows the small molecule positive and negative modulators with differential $K_{Ca}2$ subtype selectivity [44]. For treating CF and other mucociliary diseases, $K_{Ca}3.1$ inhibitors are needed [5]. Senicapoc [95] and TRAM-34 [96] inhibit $K_{Ca}3.1$ channels with IC_{50} values of ~11 nM, and ~20 nM, respectively, and they are highly selective for $K_{Ca}3.1$ channels over $K_{Ca}2$ channel subtypes [33]. The selective negative modulator for the $K_{Ca}2$ channel AP14145 is equipotent in inhibiting $K_{Ca}2.2$ and $K_{Ca}2.3$ but is ineffective on $K_{Ca}3.1$ channels [97].

For treating anosmia, COPD and its related pulmonary hypertension, $K_{Ca}2$-positive modulators may be beneficial [42,46,62]. NS309 is a potent, non-selective activator of human $K_{Ca}3.1$ and $K_{Ca}2$ channels [98]. The $K_{Ca}2.2$ and $K_{Ca}2.3$ channels are potently and selectively activated by CyPPA [38], and their derivatives are chemically modified to create more efficient and selective positive modulators [99]. However, further investigations are needed to determine their effectiveness [33].

Table 3. Small-molecule positive and negative modulators of $K_{Ca}2$ and $K_{Ca}3.1$ channels.

	Nonselective $K_{Ca}2/K_{Ca}3.1$	$K_{Ca}2$ Selective	$K_{Ca}3.1$ Selective	Subtype $K_{Ca}2$ Selective
				$K_{Ca}2.2/K_{Ca}2.3$ selective
Positive modulators	NS309 [98] SKA-31 [100] 1-EBIO [101] Riluzole [102]		SKA-111 [44] SKA-121 [103]	CyPPA [38] NS13001 [104] Compound 2q * [99]
				$K_{Ca}2.1$ selective
				CM-TPMF [102]
Negative modulators	RA-2 [103]	NS5893 [104] AP14145 [97]	Senicapoc [11,95] TRAM-34 [96]	**$K_{Ca}2.1$ selective** Bu-TPMF [102]

* 2q is a CyPPA-modified compound, other CyPPA modified compounds include: 2m–2n, 2p, 2r–2t, 2v, and 4. The potencies of these compounds on potentiating $K_{Ca}2.3$ and $K_{Ca}2.2a$ channels have previously been determined [57,99].

6. Conclusions and Perspectives

In recent years, remarkable progress has been made in understanding the physiological and pathophysiological roles of K_{Ca} channels. The advances in understanding the $K_{Ca}3.1$ structure and the resulting improvements in other $K_{Ca}2$ subtypes modeling have yet to be used, not only for drug discovery but also for understanding the pathophysiological diseases, particularly airway diseases, and developing more subtype-selective biophysical and pharmacological tools. Over the past few years, researchers have studied $K_{Ca}3.1$ channel expression and its physiological role in airway diseases. There are, however, few studies on $K_{Ca}2$ channels in the respiratory system. Evidence now suggests that $K_{Ca}2$ channels are present in the respiratory system and play an important role in airway disorders, such as asthma, chronic obstructive pulmonary disease, cystic fibrosis, and other muco-obstructive diseases. Nevertheless, further studies are necessary to unveil the exact cell distribution, subcellular localization, and protein interactions of $K_{Ca}2$ channels in the airways. Additional research is required to further establish and validate $K_{Ca}2$ and $K_{Ca}3.1$ channels as ion channels in airway diseases, their clinical relevance, and the development of more potent and subtype selective $K_{Ca}2$ channel modulators.

Author Contributions: R.O. drafted the manuscript; A.A., M.A.R. and L.L. edited and revised the manuscript; Y.-W.N. and M.Z. approved the final version of the manuscript. All authors contributed to the manuscript and the figures. All authors have read and agreed to the published version of the manuscript.

Funding: This study was supported by a 23AIREA1039423 grant from the American Heart Association, a YI-SCA grant from the National Ataxia Foundation, and a 4R33NS101182-03 grant from NIH awarded to M.Z. And grant 023-012 from King Fahad Medical City Research Center awarded to R.O.

Institutional Review Board Statement: Not applicable.

Informed Consent Statement: Not applicable.

Data Availability Statement: Not applicable.

Acknowledgments: We thank King Fahad Medical City Research Center and Chapman University Writing Center for revising the manuscript. We thank Fan Feng and Reem Orfali for their technical support. Figures were created with BioRender and published with permission.

Conflicts of Interest: The authors declare no conflict of interest.

Abbreviations

ASL	Airway surface liquid
Ca^{2+}	Calcium
CaM	Calmodulin
Cl^-	Chloride
COPD	Chronic obstructive pulmonary disease
CBF	Cilia beating frequency
K_{Ca}	Ca^{2+}-activated K^+ channels
CF	Cystic fibrosis
CFTR	Cystic fibrosis transmembrane conductance regulator
$K_{Ca}3.1$	Intermediate-conductance Ca^{2+}-activated K^+ channels
BK	Large-conductance Ca^{+2}-activated K^+ channels
MCC	Mucociliary clearance
ORNs	Olfactory receptor neurons
K^+	Potassium
$K_{Ca}2$	Small-conductance Ca^{2+}-activated K^+ channels
Na^+	Sodium
TMs	Transmembrane helices
WT	Wild type

References

1. Bartoszewski, R.; Matalon, S.; Collawn, J.F. Ion channels of the lung and their role in disease pathogenesis. *Am. J. Physiol. Cell. Mol. Physiol.* **2017**, *313*, L859–L872. [CrossRef] [PubMed]
2. Mall, M.A. Role of cilia, mucus, and airway surface liquid in mucociliary dysfunction: Lessons from mouse models. *J. Aerosol. Med. Pulm. Drug Deliv.* **2008**, *21*, 13–24. [CrossRef] [PubMed]
3. Munkholm, M.; Mortensen, J. Mucociliary clearance: Pathophysiological aspects. *Clin. Physiol. Funct. Imaging* **2014**, *34*, 171–177. [CrossRef]
4. Bhowmik, A.; Chahal, K.; Austin, G.; Chakravorty, I. Improving mucociliary clearance in chronic obstructive pulmonary disease. *Respir. Med.* **2009**, *103*, 496–502. [CrossRef]
5. Vega, G.; Guequén, A.; Philp, A.R.; Gianotti, A.; Arzola, L.; Villalón, M.; Zegarra-Moran, O.; Galietta, L.J.; Mall, M.A.; Flores, C.A. Lack of Kcnn4 improves mucociliary clearance in muco-obstructive lung disease. *JCI Insight* **2020**, *5*, 140076. [CrossRef] [PubMed]
6. Roy, M.G.; Livraghi-Butrico, A.; Fletcher, A.A.; McElwee, M.M.; Evans, S.E.; Boerner, R.M.; Alexander, S.N.; Bellinghausen, L.K.; Song, A.S.; Petrova, Y.M.; et al. Muc5b is required for airway defence. *Nature* **2014**, *505*, 412–416. [CrossRef]

7. Spina, D. Epithelium smooth muscle regulation and interactions. *Am. J. Respir. Crit. Care Med.* **1998**, *158*, S141–S145. [CrossRef]
8. Callaghan, P.J.; Ferrick, B.; Rybakovsky, E.; Thomas, S.; Mullin, J.M. Epithelial barrier function properties of the 16HBE14o- human bronchial epithelial cell culture model. *Biosci. Rep.* **2020**, *40*, BSR20201532. [CrossRef]
9. Hollenhorst, M.I.; Richter, K.; Fronius, M. Ion transport by pulmonary epithelia. *J. Biomed. Biotechnol.* **2011**, *2011*, 174306. [CrossRef]
10. Bernard, K.; Bogliolo, S.; Soriani, O.; Ehrenfeld, J. Modulation of calcium-dependent chloride secretion by basolateral SK4-like channels in a human bronchial cell line. *J. Membr. Biol.* **2003**, *196*, 15–31. [CrossRef]
11. Staal, R.G.W.; Weinstein, J.R.; Nattini, M.; Cajina, M.; Chandresana, G.; Möller, T. Senicapoc: Repurposing a Drug to Target Microglia $K_{Ca}3.1$ in Stroke. *Neurochem. Res.* **2017**, *42*, 2639–2645. [CrossRef]
12. Bardou, O.; Trinh, N.T.N.; Brochiero, E. Molecular diversity and function of K^+ channels in airway and alveolar epithelial cells. *Am. J. Physiol. Lung Cell. Mol. Physiol.* **2009**, *296*, L145–L155. [CrossRef] [PubMed]
13. KCa3.1 Channel Blockade Attenuates Microvascular Remodelling in a Large Animal Model of Bleomycin-Induced Pulmonary Fibrosis | Scientific Reports n.d. Available online: https://www.nature.com/articles/s41598-019-56412-z (accessed on 15 March 2023).
14. Zühlke, R.D.; Pitt, G.S.; Deisseroth, K.; Tsien, R.W.; Reuter, H. Calmodulin supports both inactivation and facilitation of L-type calcium channels. *Nature* **1999**, *399*, 159–162. [CrossRef] [PubMed]
15. Nam, Y.-W.; Orfali, R.; Liu, T.; Yu, K.; Cui, M.; Wulff, H.; Zhang, M. Structural insights into the potency of SK channel positive modulators. *Sci. Rep.* **2017**, *7*, 17178. [CrossRef]
16. [PDF] Unstructured to Structured Transition of an Intrinsically Disordered Protein Peptide in Coupling Ca2+-Sensing and SK Channel Activation | Semantic Scholar n.d. Available online: https://www.semanticscholar.org/paper/Unstructured-to-structured-transition-of-an-protein-Zhang-Pascal/9f5cff1624625aaaeb67cab5639813dba6db10b4 (accessed on 15 March 2023).
17. Catacuzzeno, L.; Fioretti, B.; Franciolini, F. Expression and Role of the Intermediate-Conductance Calcium-Activated Potassium Channel KCa3.1 in Glioblastoma. *J. Signal. Transduct.* **2012**, *2012*, 421564. [CrossRef] [PubMed]
18. Jensen, B.S.; Strøbaek, D.; Olesen, S.P.; Christophersen, P. The $Ca2^+$-activated K^+ channel of intermediate conductance: A molecular target for novel treatments? *Curr. Drug. Targets* **2001**, *2*, 401–422. [CrossRef]
19. Kaczmarek, L.K.; Aldrich, R.W.; Chandy, K.G.; Grissmer, S.; Wei, A.D.; Wulff, H. International Union of Basic and Clinical Pharmacology. C. Nomenclature and Properties of Calcium-Activated and Sodium-Activated Potassium Channels. *Pharmacol. Rev.* **2017**, *69*, 1–11. [CrossRef]
20. Christophersen, P.; Wulff, H. Pharmacological gating modulation of small- and intermediate-conductance Ca^{2+} -activated K^+ channels ($K_{Ca}2$.x and $K_{Ca}3.1$). *Channels* **2015**, *9*, 336–343. [CrossRef]
21. Padula, A.E.; Griffin, W.C.; Lopez, M.F.; Nimitvilai, S.; Cannady, R.; McGuier, N.S.; Chesler, E.J.; Miles, M.F.; Williams, R.W.; Randall, P.K.; et al. KCNN Genes that Encode Small-Conductance $Ca2^+$-Activated K^+ Channels Influence Alcohol and Drug Addiction. *Neuropsychopharmacol* **2015**, *40*, 1928–1939. [CrossRef] [PubMed]
22. Begenisich, T.; Nakamoto, T.; Ovitt, C.E.; Nehrke, K.; Brugnara, C.; Alper, S.L.; Melvin, J.E. Physiological Roles of the Intermediate Conductance, $Ca2^+$-activated Potassium Channel Kcnn4*. *J. Biol. Chem.* **2004**, *279*, 47681–47687. [CrossRef]
23. Orfali, R.; Albanyan, N. $Ca2^+$-Sensitive Potassium Channels. *Molecules* **2023**, *28*, 885. [CrossRef] [PubMed]
24. Girault, A.; Haelters, J.-P.; Potier-Cartereau, M.; Chantôme, A.; Jaffrés, P.-A.; Bougnoux, P.; Joulin, V.; Vandier, C. Targeting SKCa channels in cancer: Potential new therapeutic approaches. *Curr. Med. Chem.* **2012**, *19*, 697–713. [CrossRef]
25. Aldrich, R.; Chandy, K.G.; Grissmer, S.; Gutman, G.A.; Kaczmarek, L.K.; Wei, A.D.; Wulff, H. Calcium- and sodium-activated potassium channels (K_{Ca}, K_{Na}) in GtoPdb v.2021.3. *GtoPdb CITE* **2021**, *2021*. [CrossRef]
26. Xia, X.-M.; Fakler, B.; Rivard, A.; Wayman, G.; Johnson-Pais, T.; Keen, J.E.; Ishii, T.; Hirschberg, B.; Bond, C.T.; Lutsenko, S.; et al. Mechanism of calcium gating in small-conductance calcium-activated potassium channels. *Nature* **1998**, *395*, 503–507. [CrossRef] [PubMed]
27. Shmukler, B.E.; Bond, C.T.; Wilhelm, S.; Bruening-Wright, A.; Maylie, J.; Adelman, J.P.; Alper, S.L. Structure and complex transcription pattern of the mouse SK1 K_{Ca} channel gene, KCNN1. *Biochim. Biophys. Acta* **2001**, *1518*, 36–46. [CrossRef] [PubMed]
28. Zhang, B.M.; Kohli, V.; Adachi, R.; López, J.A.; Udden, M.M.; Sullivan, R. Calmodulin binding to the C-terminus of the small-conductance Ca^{2+}-activated K^+ channel hSK1 is affected by alternative splicing. *Biochemistry* **2001**, *40*, 3189–3195. [CrossRef]
29. Nam, Y.-W.; Cui, M.; El-Sayed, N.S.; Orfali, R.; Nguyen, M.; Yang, G.; Rahman, M.A.; Lee, J.; Zhang, M. Subtype-selective positive modulation of KCa2 channels depends on the HA/HB helices. *Br. J. Pharmacol.* **2022**, *179*, 460–472. [CrossRef] [PubMed]
30. Nam, Y.-W.; Cui, M.; Orfali, R.; Viegas, A.; Nguyen, M.; Mohammed, E.H.M.; Zoghebi, K.A.; Rahighi, S.; Parang, K.; Zhang, M. Hydrophobic interactions between the HA helix and S4-S5 linker modulate apparent Ca^{2+} sensitivity of SK2 channels. *Acta Physiol.* **2021**, *231*, e13552. [CrossRef]
31. El-Sayed, N.S.; Nam, Y.-W.; Egorova, P.A.; Nguyen, H.M.; Orfali, R.; Rahman, M.A.; Yang, G.; Wulff, H.; Bezprozvanny, I.; Parang, K.; et al. Structure-Activity Relationship Study of Subtype-Selective Positive Modulators of $K_{Ca}2$ Channels. *J. Med. Chem.* **2022**, *65*, 303–322. [CrossRef]
32. Morales, P.; Garneau, L.; Klein, H.; Lavoie, M.-F.; Parent, L.; Sauvé, R. Contribution of the $K_{Ca}3.1$ channel–calmodulin interactions to the regulation of the $K_{Ca}3.1$ gating process. *J. Gen. Physiol.* **2013**, *142*, 37–60. [CrossRef] [PubMed]
33. Nam, Y.-W.; Downey, M.; Rahman, M.A.; Cui, M.; Zhang, M. Channelopathy of small- and intermediate-conductance Ca^{2+}-activated K^+ channels. *Acta Pharmacol. Sin.* **2023**, *44*, 259–267. [CrossRef]

34. Orfali, R.; Nam, Y.-W.; Nguyen, H.M.; Rahman, M.A.; Yang, G.; Cui, M.; Wulff, H.; Zhang, M. Channelopathy-causing mutations in the S45A/S45B and HA/HB helices of $K_{Ca}2.3$ and $K_{Ca}3.1$ channels alter their apparent Ca^{2+} sensitivity. *Cell Calcium* **2022**, *102*, 102538. [CrossRef] [PubMed]
35. Monaghan, A.S.; Benton, D.C.H.; Bahia, P.K.; Hosseini, R.; Shah, Y.A.; Haylett, D.G.; Moss, G.W.J. The SK3 Subunit of Small Conductance Ca^{2+}-activated K^{+} Channels Interacts with Both SK1 and SK2 Subunits in a Heterologous Expression System. *J. Biol. Chem.* **2004**, *279*, 1003–1009. [CrossRef] [PubMed]
36. Lee, C.-H.; MacKinnon, R. Activation mechanism of a human SK-calmodulin channel complex elucidated by cryo-EM structures. *Science* **2018**, *360*, 508–513. [CrossRef] [PubMed]
37. Ishii, T.M.; Silvia, C.; Hirschberg, B.; Bond, C.T.; Adelman, J.P.; Maylie, J. A human intermediate conductance calcium-activated potassium channel. *Proc. Natl. Acad. Sci. USA* **1997**, *94*, 11651–11656. [CrossRef] [PubMed]
38. Hougaard, C.; Eriksen, B.L.; Jørgensen, S.; Johansen, T.H.; Dyhring, T.; Madsen, L.S.; Strøbaek, D.; Christophersen, P. Selective positive modulation of the SK3 and SK2 subtypes of small conductance Ca^{2+}-activated K^{+} channels. *Br. J. Pharmacol.* **2007**, *151*, 655–665. [CrossRef] [PubMed]
39. Orfali, R. Genetic Mutations of $K_{Ca}2.3$ and $K_{Ca}3.1$ Channels Affect Ca^{2+} Sensitivity. Ph.D. Thesis, Pharmaceutical Sciences, Chapman University, Irvine, CA, USA, 2023. [CrossRef]
40. Wulff, H.; Zhorov, B.S. K^{+} Channel Modulators for the Treatment of Neurological Disorders and Autoimmune Diseases. *Chem. Rev.* **2008**, *108*, 1744–1773. [CrossRef] [PubMed]
41. Bardou, O.; Trinh, N.T.N.; Brochiero, E. Canaux potassiques et physiologie de l'épithélium respiratoire. *Med. Sci.* **2009**, *25*, 391–397. [CrossRef]
42. Kroigaard, C.; Dalsgaard, T.; Nielsen, G.; Laursen, B.E.; Pilegaard, H.; Köhler, R.; Simonsen, U. Activation of endothelial and epithelial $K_{Ca}2.3$ calcium-activated potassium channels by NS309 relaxes human small pulmonary arteries and bronchioles. *Br. J. Pharmacol.* **2012**, *167*, 37–47. [CrossRef] [PubMed]
43. Rahm, A.; Wieder, T.; Gramlich, D.; Müller, M.E.; Wunsch, M.N.; El Tahry, F.A.; Heimberger, T.; Sandke, S.; Weis, T.; Most, P.; et al. Differential regulation of $K_{Ca}2.1$ (KCNN1) K^{+} channel expression by histone deacetylases in atrial fibrillation with concomitant heart failure. *Physiol. Rep.* **2021**, *9*, e14835. [CrossRef]
44. Brown, B.M.; Shim, H.; Christophersen, P.; Wulff, H. Pharmacology of Small- and Intermediate-Conductance Calcium-Activated Potassium Channels. *Annu. Rev. Pharmacol. Toxicol.* **2020**, *60*, 219–240. [CrossRef] [PubMed]
45. Chen, M.X.; Gorman, S.A.; Benson, B.; Singh, K.; Hieble, J.P.; Michel, M.C.; Tate, S.N.; Trezise, D.J. Small and intermediate conductance Ca^{2+}-activated K^{+} channels confer distinctive patterns of distribution in human tissues and differential cellular localisation in the colon and corpus cavernosum. *Naunyn-Schmiedeberg's Arch. Pharmacol.* **2004**, *369*, 602–615. [CrossRef] [PubMed]
46. Wulff, H.; Köhler, R. Endothelial Small- and Intermediate-Conductance KCa Channels: An Update on Their Pharmacology and Usefulness as Cardiovascular Targets. *J. Cardiovasc. Pharmacol.* **2013**, *61*, 102–112. [CrossRef]
47. Schumacher, M.A.; Rivard, A.F.; Bächinger, H.P.; Adelman, J.P. Structure of the gating domain of a Ca^{2+}-activated K^{+} channel complexed with Ca^{2+}/calmodulin. *Nature* **2001**, *410*, 1120–1124. [CrossRef]
48. Zhang, M.; Abrams, C.; Wang, L.; Gizzi, A.; He, L.; Lin, R.; Chen, Y.; Loll, P.J.; Pascal, J.M.; Zhang, J. Structural basis for calmodulin as a dynamic calcium sensor. *Structure* **2012**, *20*, 911–923. [CrossRef]
49. Nguyen, T.N.; Suzuki, H.; Ohkubo, J.-I.; Wakasugi, T.; Kitamura, T. Calmodulin Regulates Ciliary Beats in the Human Nasal Mucosa Through Adenylate/Guanylate Cyclases and Protein Kinases A/G. *Int. Arch. Allergy Immunol.* **2021**, *182*, 800–806. [CrossRef] [PubMed]
50. Valeyev, N.V.; Bates, D.G.; Heslop-Harrison, P.; Postlethwaite, I.; Kotov, N.V. Elucidating the mechanisms of cooperative calcium-calmodulin interactions: A structural systems biology approach. *BMC Syst. Biol.* **2008**, *2*, 48. [CrossRef] [PubMed]
51. Restrepo, D.; Okada, Y.; Teeter, J.H.; Lowry, L.D.; Cowart, B.; Brand, J.G. Human olfactory neurons respond to odor stimuli with an increase in cytoplasmic Ca^{2+}. *Biophys. J.* **1993**, *64*, 1961–1966.
52. Delgado, R.; Saavedra, M.V.; Schmachtenberg, O.; Sierralta, J.; Bacigalupo, J. Presence of Ca^{2+}-dependent K^{+} channels in chemosensory cilia support a role in odor transduction. *J. Neurophysiol.* **2003**, *90*, 2022–2028. [CrossRef] [PubMed]
53. Di Benedetto, G.; Magnus, C.J.; Gray, P.T.; Mehta, A. Calcium regulation of ciliary beat frequency in human respiratory epithelium in vitro. *J. Physiol.* **1991**, *439*, 103–113. [CrossRef]
54. De Proost, I.; Pintelon, I.; Brouns, I.; Kroese, A.B.A.; Riccardi, D.; Kemp, P.J.; Timmermans, J.-P.; Adriaensen, D. Functional live cell imaging of the pulmonary neuroepithelial body microenvironment. *Am. J. Respir. Cell Mol. Biol.* **2008**, *39*, 180–189. [CrossRef] [PubMed]
55. Ohba, T.; Sawada, E.; Suzuki, Y.; Yamamura, H.; Ohya, S.; Tsuda, H.; Imaizumi, Y. Enhancement of Ca^{2+} influx and ciliary beating by membrane hyperpolarization due to ATP-sensitive K^{+} channel opening in mouse airway epithelial cells. *J. Pharmacol. Exp. Ther.* **2013**, *347*, 145–153. [CrossRef]
56. Tilley, A.E.; Walters, M.S.; Shaykhiev, R.; Crystal, R.G. Cilia Dysfunction in Lung Disease. *Annu. Rev. Physiol.* **2015**, *77*, 379–406. [CrossRef]
57. Hebert, S.C.; Desir, G.; Giebisch, G.; Wang, W. Molecular diversity and regulation of renal potassium channels. *Physiol. Rev.* **2005**, *85*, 319–371. [CrossRef] [PubMed]

58. Nam, Y.-W.; Pala, R.; El-Sayed, N.S.; Larin-Henriquez, D.; Amirrad, F.; Yang, G.; Rahman, M.A.; Orfali, R.; Downey, M.; Parang, K.; et al. Subtype-Selective Positive Modulation of K_{Ca}2.3 Channels Increases Cilia Length. *ACS Chem. Biol.* **2022**, *17*, 2344–2354. [CrossRef]
59. Weaver, A.K.; Bomben, V.C.; Sontheimer, H. Expression and Function of Calcium-Activated Potassium Channels in Human Glioma Cells. *Glia* **2006**, *54*, 223–233. [CrossRef] [PubMed]
60. Stocker, M. Ca(2+)-activated K+ channels: Molecular determinants and function of the SK family. *Nat. Rev. Neurosci.* **2004**, *5*, 758–770. [CrossRef]
61. Shepherd, M.C.; Duffy, S.M.; Harris, T.; Cruse, G.; Schuliga, M.; Brightling, C.E.; Neylon, C.B.; Bradding, P.; Stewart, A.G. K_{Ca}3.1 Ca^{2+}Activated K^+ Channels Regulate Human Airway Smooth Muscle Proliferation. *Am. J. Respir. Cell Mol. Biol.* **2007**, *37*, 525–531. [CrossRef]
62. Kunzelmann, K.; Pavenstädt, H.; Beck, C.; Ünal, Ö.; Emmrich, P.; Arndt, H.J.; Greger, R. Characterization of potassium channels in respiratory cells. I. General properties. *Pflugers Arch.* **1989**, *414*, 291–296. [CrossRef]
63. Pablo, J.L.; DeCaen, P.G.; Clapham, D.E. Progress in ciliary ion channel physiology. *J. Gen. Physiol.* **2017**, *149*, 37–47. [CrossRef]
64. Kawai, F. Ca^{2+}-activated K^+ currents regulate odor adaptation by modulating spike encoding of olfactory receptor cells. *Biophys. J.* **2002**, *82*, 2005–2015. [CrossRef] [PubMed]
65. Kume, H.; Hall, I.P.; Washabau, R.J.; Takagi, K.; Kotlikoff, M.I. Beta-Adrenergic Agonists Regulate KCa Channels in Airway Smooth Muscle by cAMP-Dependent and -Independent Mechanisms. 1994. Available online: https://www.jci.org/articles/view/116969/scanned-page/371 (accessed on 17 March 2023).
66. Fujita, A.; Takeuchi, T.; Hanai, J.; Hata, F. Expression of the small conductance Ca^{2+}-activated K^+ channel, SK3, in the olfactory ensheathing glial cells of rat brain. *Cell Tissue Res.* **2003**, *313*, 187–193. [CrossRef] [PubMed]
67. Collawn, J.F.; Matalon, S. CFTR and lung homeostasis. *Am. J. Physiol. Lung Cell. Mol. Physiol.* **2014**, *307*, L917–L923. [CrossRef]
68. Naeije, R.; Barberà, J.A. Pulmonary hypertension associated with COPD. *Crit. Care* **2001**, *5*, 286–289. [CrossRef] [PubMed]
69. López-Barneo, J.; del Toro, R.; Levitsky, K.L.; Chiara, M.D.; Ortega-Sáenz, P. Regulation of oxygen sensing by ion channels. *J. Appl. Physiol.* **2004**, *96*, 1187–1195. [CrossRef]
70. Franco-Obregón, A.; López-Barneo, J. Low PO2 inhibits calcium channel activity in arterial smooth muscle cells. *Am. J. Physiol.* **1996**, *271*, H2290–H2299. [CrossRef]
71. Franco-Obregón, A.; Ureña, J.; López-Barneo, J. Oxygen-sensitive calcium channels in vascular smooth muscle and their possible role in hypoxic arterial relaxation. *Proc. Natl. Acad. Sci. USA* **1995**, *92*, 4715. [CrossRef]
72. Kume, H.; Nishiyama, O.; Isoya, T.; Higashimoto, Y.; Tohda, Y.; Noda, Y. Involvement of Allosteric Effect and KCa Channels in Crosstalk between β2-Adrenergic and Muscarinic M2 Receptors in Airway Smooth Muscle. *Int. J. Mol. Sci.* **2018**, *19*, 1999. [CrossRef] [PubMed]
73. Morales, B.; Labarca, P.; Bacigalupo, J. A ciliary K+ conductance sensitive to charibdotoxin underlies inhibitory responses in toad olfactory receptor neurons. *FEBS Lett.* **1995**, *359*, 41–44. [CrossRef]
74. Morales, B.; Madrid, R.; Bacigalupo, J. Calcium mediates the activation of the inhibitory current induced by odorants in toad olfactory receptor neurons. *FEBS Lett.* **1997**, *402*, 259–264. [CrossRef]
75. Sanhueza, M.; Schmachtenberg, O.; Bacigalupo, J. Excitation, inhibition, and suppression by odors in isolated toad and rat olfactory receptor neurons. *Am. J. Physiol. Cell Physiol.* **2000**, *279*, C31–C39. [CrossRef] [PubMed]
76. Logsdon, N.J.; Kang, J.; Togo, J.A.; Christian, E.P.; Aiyar, J. A novel gene, $hK_{Ca}4$, encodes the calcium-activated potassium channel in human T lymphocytes. *J. Biol. Chem.* **1997**, *272*, 32723–32726. [CrossRef]
77. Feske, S.; Wulff, H.; Skolnik, E.Y. Ion channels in innate and adaptive immunity. *Annu. Rev. Immunol.* **2015**, *33*, 291–353. [CrossRef] [PubMed]
78. Kaushal, V.; Koeberle, P.D.; Wang, Y.; Schlichter, L.C. The $Ca2^+$-activated K^+ channel *KCNN4*/K_{Ca}3.1 contributes to microglia activation and nitric oxide-dependent neurodegeneration. *J. Neurosci.* **2007**, *27*, 234–244. [CrossRef] [PubMed]
79. Ghanshani, S.; Wulff, H.; Miller, M.J.; Rohm, H.; Neben, A.; Gutman, G.A.; Cahalan, M.D.; Chandy, K.G. Up-regulation of the IKCa1 Potassium Channel during T-cell Activation: Molecular Mechanism and Functional Consequences. *J. Biol. Chem.* **2000**, *275*, 37137–37149. [CrossRef] [PubMed]
80. Shao, Z.; Makinde, T.O.; Agrawal, D.K. Calcium-Activated Potassium Channel K_{Ca}3.1 in Lung Dendritic Cell Migration. *Am. J. Respir. Cell Mol. Biol.* **2011**, *45*, 962–968. [CrossRef] [PubMed]
81. Tharp, D.L.; Wamhoff, B.R.; Turk, J.R.; Bowles, D.K. Upregulation of intermediate-conductance Ca^{2+}-activated K^+ channel (IKCa1) mediates phenotypic modulation of coronary smooth muscle. *Am. J. Physiol. Heart Circ. Physiol.* **2006**, *291*, H2493–H2503. [CrossRef] [PubMed]
82. Jaggar, J.H.; Wellman, G.C.; Heppner, T.J.; Porter, V.A.; Perez, G.J.; Gollasch, M.; Kleppisch, T.; Rubart, M.; Stevenson, A.S.; Lederer, W.J.; et al. Ca^{2+} channels, ryanodine receptors and Ca^{2+}-activated K^+ channels: A functional unit for regulating arterial tone. *Acta Physiol. Scand.* **1998**, *164*, 577–587. [CrossRef]
83. Neylon, C.B.; Lang, R.J.; Fu, Y.; Bobik, A.; Reinhart, P.H. Molecular cloning and characterization of the intermediate-conductance Ca^{2+}-activated K^+ channel in vascular smooth muscle: Relationship between K_{Ca} channel diversity and smooth muscle cell function. *Circ. Res.* **1999**, *85*, e33–e43. [CrossRef]
84. Peña, T.L.; Rane, S.G. The fibroblast intermediate conductance K_{Ca} channel, FIK, as a prototype for the cell growth regulatory function of the IK channel family. *J. Membr. Biol.* **1999**, *172*, 249–257. [CrossRef]

85. Gerlach, A.C.; Gangopadhyay, N.N.; Devor, D.C. Kinase-dependent regulation of the intermediate conductance, calcium-dependent potassium channel, hIK1. *J. Biol. Chem.* **2000**, *275*, 585–598. [CrossRef]
86. Wulff, H.; Castle, N.A. Therapeutic potential of $K_{Ca}3.1$ blockers: An overview of recent advances, and promising trends. *Expert Rev. Clin. Pharmacol.* **2010**, *3*, 385–396. [CrossRef]
87. Srivastava, S.; Zhdanova, O.; Di, L.; Li, Z.; Albaqumi, M.; Wulff, H.; Skolnik, E.Y. Protein histidine phosphatase 1 negatively regulates CD4 T cells by inhibiting the K^+ channel $K_{Ca}3.1$. *Proc. Natl. Acad. Sci. USA* **2008**, *105*, 14442–14446. [CrossRef] [PubMed]
88. Cruse, G.; Duffy, S.M.; Brightling, C.E.; Bradding, P. Functional $K_{Ca}3.1$ K^+ channels are required for human lung mast cell migration. *Thorax* **2006**, *61*, 880–885. [CrossRef]
89. Roach, K.M.; Duffy, S.M.; Coward, W.; Feghali-Bostwick, C.; Wulff, H.; Bradding, P. The K^+ channel $K_{Ca}3.1$ as a novel target for idiopathic pulmonary fibrosis. *PLoS ONE* **2013**, *8*, e85244. [CrossRef] [PubMed]
90. Tarran, R. Regulation of airway surface liquid volume and mucus transport by active ion transport. *Proc. Am. Thorac. Soc.* **2004**, *1*, 42–46. [CrossRef]
91. Parasuraman, S.; Walker, S.; Loudon, B.L.; Gollop, N.D.; Wilson, A.M.; Lowery, C.; Frenneaux, M.P. Assessment of pulmonary artery pressure by echocardiography—A comprehensive review. *Int. J. Cardiol. Heart Vasc.* **2016**, *12*, 45–51. [CrossRef]
92. Wulff, H.; Kolski-Andreaco, A.; Sankaranarayanan, A.; Sabatier, J.-M.; Shakkottai, V. Modulators of small- and intermediate-conductance calcium-activated potassium channels and their therapeutic indications. *Curr. Med. Chem.* **2007**, *14*, 1437–1457. [CrossRef] [PubMed]
93. Sørensen, U.S.; Strøbaek, D.; Christophersen, P.; Hougaard, C.; Jensen, M.L.; Nielsen, E.Ø.; Peters, D.; Teuber, L. Synthesis and structure-activity relationship studies of 2-(N-substituted)-aminobenzimidazoles as potent negative gating modulators ofsmall conductance Ca^{2+}-activated K^+ channels. *J. Med. Chem.* **2008**, *51*, 7625–7634. [CrossRef] [PubMed]
94. Stocker, J.W.; De Franceschi, L.; McNaughton-Smith, G.A.; Corrocher, R.; Beuzard, Y.; Brugnara, C. ICA-17043, a novel Gardos channel blocker, prevents sickled red blood cell dehydration in vitro and in vivo in SAD mice. *Blood* **2003**, *101*, 2412–2418. [CrossRef]
95. Agarwal, J.J.; Zhu, Y.; Zhang, Q.-Y.; Mongin, A.A.; Hough, L.B. TRAM-34, a putatively selective blocker of intermediate-conductance, calcium-activated potassium channels, inhibits cytochrome P450 activity. *PLoS ONE* **2013**, *8*, e63028. [CrossRef] [PubMed]
96. Simó-Vicens, R.; Kirchhoff, J.E.; Dolce, B.; Abildgaard, L.; Speerschneider, T.; Sørensen, U.S.; Grunnet, M.; Diness, J.G.; Bentzen, B.H. A new negative allosteric modulator, AP14145, for the study of small conductance calcium-activated potassium ($K_{Ca}2$) channels. *Br. J. Pharmacol.* **2017**, *174*, 4396–4408. [CrossRef]
97. Strøbaek, D.; Teuber, L.; Jørgensen, T.D.; Ahring, P.K.; Kjaer, K.; Hansen, R.S.; Olesen, S.P.; Christophersen, P.; Skaaning-Jensen, B. Activation of human IK and SK Ca^{2+} -activated K^+ channels by NS309 (6,7-dichloro-1H-indole-2,3-dione 3-oxime). *Biochim. Biophys. Acta* **2004**, *1665*, 1–5. [CrossRef]
98. John, C.M.; Khaddaj Mallat, R.; Mishra, R.C.; George, G.; Singh, V.; Turnbull, J.D.; Umeshappa, C.S.; Kendrick, D.J.; Kim, T.; Fauzi, F.M.; et al. SKA-31, an activator of Ca^{2+}-activated K^+ channels, improves cardiovascular function in aging. *Pharmacol. Res.* **2020**, *151*, 104539. [CrossRef] [PubMed]
99. The IKCa and SKCa Channel Activator 1-EBIO (100 μM) Increased Outwards .... Available online: https://www.researchgate.net/figure/The-IKCa-and-SKCa-channel-activator-1-EBIO-100-M-increased-outwards-currents-in_fig5_235718670 (accessed on 3 April 2023).
100. Dimitriadi, M.; Kye, M.J.; Kalloo, G.; Yersak, J.M.; Sahin, M.; Hart, A.C. The Neuroprotective Drug Riluzole Acts via Small Conductance Ca^{2+}-Activated K^+ Channels to Ameliorate Defects in Spinal Muscular Atrophy Models. *J. Neurosci.* **2013**, *33*, 6557–6562. [CrossRef] [PubMed]
101. Kasumu, A.W.; Hougaard, C.; Rode, F.; Jacobsen, T.A.; Sabatier, J.M.; Eriksen, B.L.; Strøbæk, D.; Liang, X.; Egorova, P.; Vorontsova, D.; et al. Selective positive modulator of calcium-activated potassium channels exerts beneficial effects in a mouse model of spinocerebellar ataxia type 2. *Chem. Biol.* **2012**, *19*, 1340–1353. [CrossRef] [PubMed]
102. Hougaard, C.; Hammami, S.; Eriksen, B.L.; Sørensen, U.S.; Jensen, M.L.; Strøbæk, D.; Christophersen, P. Evidence for a common pharmacological interaction site on $K_{Ca}2$ channels providing both selective activation and selective inhibition of the human $K_{Ca}2.1$ subtype. *Mol. Pharmacol.* **2012**, *81*, 210–219. [CrossRef]
103. Oliván-Viguera, A.; Valero, M.S.; Coleman, N.; Brown, B.M.; Laría, C.; Murillo, M.D.; Gálvez, J.A.; Díaz-de-Villegas, M.D.; Wulff, H.; Badorrey, R.; et al. A Novel Pan-Negative-Gating Modulator of $K_{Ca}2/3$ Channels, Fluoro-Di-Benzoate, RA-2, Inhibits Endothelium-Derived Hyperpolarization–Type Relaxation in Coronary Artery and Produces Bradycardia In Vivo. *Mol. Pharmacol.* **2015**, *87*, 338–348. [CrossRef] [PubMed]
104. Strøbæk, D.; Hougaard, C.; Johansen, T.H.; Sørensen, U.S.; Nielsen, E.Ø.; Nielsen, K.S.; Taylor, R.D.T.; Pedarzani, P.; Christophersen, P. Inhibitory Gating Modulation of Small Conductance Ca^{2+}-Activated K^+ Channels by the Synthetic Compound (R)-N-(Benzimidazol-2-yl)-1,2,3,4-tetrahydro-1-naphtylamine (NS8593) Reduces Afterhyperpolarizing Current in Hippocampal CA1 Neurons. *Mol. Pharmacol.* **2006**, *70*, 1771–1782. [CrossRef] [PubMed]

biomedicines

Review

Better Safe than Sorry: Rheumatoid Arthritis, Interstitial Lung Disease, and Medication—A Narrative Review

Iulia-Tania Andronache [1,2,*], Victoria-Cristina Şuţa [3], Maria Şuţa [1], Sabina-Livia Ciocodei [1], Liliana Vladareanu [1], Alina Doina Nicoara [3] and Oana Cristina Arghir [1,4]

1 Doctoral School of Medicine, "Ovidius" University of Constanta, 900470 Constanta, Romania; maria_suta@yahoo.com (M.Ş.); arghir_oana@yahoo.com (O.C.A.)
2 Department of Rheumatology, Internal Medicine Clinic, "Dr. Alexandru Gafencu" Military Emergency Hospital Constanta, 900527 Constanta, Romania
3 3rd Department—1st Clinical Medical Disciplines, Faculty of Medicine, "Ovidius" University of Constanta, 900470 Constanta, Romania
4 4th Department—2nd Clinical Medical Disciplines, Faculty of Medicine, "Ovidius" University of Constanta, 900470 Constanta, Romania
* Correspondence: andronacheiulia@gmail.com

Abstract: It is well known that rheumatoid arthritis (RA) patients are at an increased risk of developing non-infectious pulmonary complications, especially interstitial lung disease (ILD); however, the clinician must keep in mind that lung disease could not only be a manifestation of the underlying condition, but also a consequence of using disease-modifying therapies. New-onset ILD or ILD worsening has also been reported as a possible consequence of both conventional disease-modifying antirheumatic drugs (DMARDs) and biologic agents. This study is a narrative review of the current literature regarding the potential risk of developing interstitial lung disease along with the administration of specific drugs used in controlling rheumatoid arthritis. Its purpose is to fill knowledge gaps related to this challenging patient cohort by addressing various aspects of the disease, including prevalence, disease features, treatment strategies, and patient outcomes.

Keywords: rheumatoid arthritis; drug-induced lung injury; interstitial lung disease; pulmonary fibrosis

1. Introduction

Drug-induced lung injury (DLI) is defined as a lung injury that specifically results from the use of a drug, including prescription and over-the-counter drugs, herbal medicines, supplements, and illegal narcotics [1].

As causality is often difficult to prove, attributing lung injury to medication is a challenging task. The diagnosis is usually presumptive, lacking a gold standard test, and the causal connection is determined by the temporal association between drug initiation, symptom onset, radiologic abnormalities, the failure to identify a microbial agent, the absence of pre-existing lung disease, and improvement after drug discontinuation [1,2].

Based on these principles, the 2013 Consensus statement for the diagnosis and treatment of drug-induced lung injuries recommends the following criteria for drug-induced lung injury [1,3,4] (Table 1).

Although the exact pathogenic mechanisms of drug-induced lung injury (DLI) have not been fully understood, except for a limited number of drugs, two potential primary mechanisms have been proposed: the cytotoxic and the immune mechanisms of action. These mechanisms may be independently involved or in combination, resulting in various forms of lung injury. These two mechanisms may be influenced by contextual factors, such as smoking or underlying lung diseases (Table 2) [1].

Table 1. Diagnostic criteria for drug-induced lung injury [3].

History of Ingestion of a Drug that is Known to Induce Lung Injury	Specifically Inquire about the Following when Taking the Patient's History: Over-the-Counter (OTC) Drugs, Health Foods, and Illegal Narcotic Drugs/Anti-Hypnotic Drugs
The clinical manifestation has been reported to be induced by a drug	The clinical manifestations include clinical findings, imaging findings, and pathological features.
Other causes of the clinical manifestation could be ruled out	Differentiation from infection, cardiogenic pulmonary edema, exacerbation of an underlying disease, etc.
Improvement of the clinical manifestations after drug discontinuation	Spontaneous remission or remission in response to a corticosteroid.
Exacerbation of the clinical manifestations after resuming drug administration	Resuming drug administration to identify the causative drug is not generally recommended but is acceptable if the patient requires the drug and safety is assured.

Table 2. Pathogenetic mechanisms of drug-induced lung injury.

Cytotoxic effects (e.g., methotrexate, cyclophosphamide, sulfasalazine)	Direct damage to alveolar epithelium and endothelialium. Mediated by reactive oxygen species, proteases, and cytokines. Pulmonary fibrosis due to increased vascular permeability, inflammation, and tissue injury (extent of injury is generally usually dose and duration dependent).
Activation of immune cells (allergic reaction) (e.g., anti TNF alpha)	Immunogenicity due to drug/drugmetabolite binding to cytoplasmic proteins (hapten hypothesis). Eosinophil infiltration in the alveolar wall and airspaces (eosinophilic pneumonia). Interstitial pneumonia mediated by lymphocyte infiltration and granuloma formation in the alveolar wall (involved in most cases of DLI and usually neither dose nor duration related).
Host factors	Age, smoker status, exposure to chemicals/dusts, genetic factors. Underlying pulmonary diseases: pulmonary fibrosis, chronic obstructive pulmonary disease (COPD), and emphysema. Iatrogenic: exposure to high concentrations of oxygen, history of radiation exposure, surgery

Adapted from [1].

Whilst the use of biologic disease-modifying agents to treat RA has led to an unprecedented improvement in clinical and functional outcomes, their safety profile remains to be fully elucidated. Almost all DMARDs are known to potentially trigger DLI in patients with RA [1,5,6]. A very useful tool when considering DLI is The Drug-Induced Respiratory Disease Website developed by Philippe Camus—http://www.pneumotox.com (accessed on 10 April 2023). This website is a simple, easy-to-use tool, in which, after introducing the generic name of the drug, a frequency or incidence gauge (in the form of digits within stars) indicates how many cases have been published in the literature (questionable signal; 1 <10 cases; 2 10–50 cases; 3 50–100 cases; 4 100–200 cases; 5 >200 cases) [5]. The website also provides abundant links to the existing literature and case reports.

The treat-to-target (T2T) strategy in rheumatoid arthritis (RA) is an approach that involves setting a treatment goal (RA remission or low disease activity) based on disease activity scores [7]. This strategy involves early aggressive treatment with one or more conventional synthetic DMARDs and/or biologic DMARDs/targeted—tsDMARDs (Table 3), along with symptomatic therapy that may include NSAIDs, low-dose prednisone, and physical therapy.

Table 3. DMARDs used in RA treatment [8].

Synthetic DMARDs	Conventional—csDMARDs	Methotrexate, Leflunomide, Hydroxychloroquine, Sulfasalazine, Cyclosporine, Azathioprine
	Targeted—tsDMARDs/JAK Inhibitors	Tofacitinib, Baricitinib, Upadacitinib,
Biological DMARDs	TNF Alpha inhibitors	Infliximab, Etanercept, Adalimumab, Golimumab, Certolizumab
	CD 20 receptors on B cells inhibitor	Rituximab
	Interleukin 6 receptor antagonist	Tocilizumab
	Selective T cell co-stimulation modulator	Abatacept
	Interleukin 1 receptor antagonist	Anakinra

2. Conventional DMARDS

2.1. Methotrexate

By inhibiting the enzyme AICAR transformylase, Methotrexate disrupts the metabolism of adenosine and guanine, resulting in the accumulation of adenosine. The increased levels of adenosine contribute to the anti-inflammatory effects by suppressing T-cell activation, down-regulating B-cells, and enhancing the sensitivity of activated CD-95 T cells. [5,8]

In retrospective studies, the incidence of Methotrexate-induced lung disease has been estimated to be 3.5–7.6%, with a prevalence of 5% [9], and has been shown to develop early during Methotrexate treatment, mostly within the first year [10].

Methotrexate can lead to multiple respiratory problems: hypersensitivity pneumonitis, interstitial fibrosis, acute lung injury with noncardiogenic pulmonary edema, organizing pneumonia, pleuritis and pleural effusions, and pulmonary nodules with acute interstitial hypersensitivity pneumonitis being the most common [11,12].

Methotrexate-induced pneumonitis is reported in about 0.43% to 1% of treated patients [10] with a mortality rate of 20% [13].

The risk factors for methotrexate-induced lung injury in RA patients are age (>60 years), previous DMARDs use (LEF, SSZ, gold, or D-penicillamine), preexistent respiratory manifestations of RA, hypoalbuminemia, diabetes, and renal dysfunction [11].

Sathi et al. adds as an additional risk factor for Methotrexate pneumonitis reduced baseline pulmonary function tests (PFT), advising baseline pulmonary function tests when initiating Methotrexate therapy [14]. However, the role of Methotrexate in the development of ILD in RA patients is debatable since the severity of RA is an independent risk factor for developing ILD [9,14].

In a meta-analysis of randomized controlled trials, Conway et al. demonstrated a slightly increased risk of respiratory adverse events, including ILD exacerbations in patients with RA treated with Methotrexate compared with other DMARDs and biologic agents (RR 1.10, 95% CI 1.02–1.19). However, the patients treated with Methotrexate did not have an increased risk of death due to pulmonary disease [9].

Methotrexate pneumonitis typically has an acute/subacute onset and often presents a hypersensitivity pneumonitis pattern [15]. Several days/weeks after initiating Methotrexate (or more gradually in the subacute form), the patient presents dyspnea, non-productive cough, fever, and, in some cases, acute respiratory failure [15,16].

Allergic mechanisms are thought to cause DLI because respiratory symptoms are often accompanied by fever and peripheral eosinophilia. Drug-induced lung injury most frequently shows widespread ground glass opacity (GGO) and hypersensitivity pneumonitis-like pattern, which may resemble the imaging findings from pneumocystis pneumonia (PCP) [15].

For a better diagnostic approach there have been developed diagnosis criteria for adverse pulmonary events associated with Methrotrexate use (Table 4).

Table 4. Searles and McKendry diagnostic criteria for adverse pulmonary events associated with methotrexate treatment in rheumatoid arthritis [16,17].

Major criteria	I. Hypersensitivity pneumonitis (demonstrated by histopathologic examination, with no evidence of pathogenic organisms).
	2. Radiologic evidence of pulmonary interstitial or alveolar infiltrates.
	3. Negative blood (if afebrile) and initial sputum (if productive) cultures.
Minor criteria	1. Shortness of breath of <8 weeks duration.
	2. Non-productive cough.
	3. O2 saturation < 90% at the time of initial evaluation.
	4. DLCO < 70% of the predicted value
	5. WBC $\leq$ 15,000 per mm^3.

Definite cases were defined as the presence of major criterion 1 or major criteria 2 and 3, and three out of the five minor criteria. Probable cases were defined as the presence of major criteria 2 and 3, and two minor criteria. DLCO = diffusing capacity for carbon monoxide; WBC = white blood cell (count).

Adapted from [16,17].

The topic of chronic pulmonary injury induced by Methotrexate is highly debated: when investigating the chronic pulmonary effects of low-dose Methotrexate in RA patients (55 RA patients treated with Methotrexate and 75 RA patients in the control group), Dawson et al. did not find any evidence that Methotrexate leads to chronic pulmonary fibrosis. Moreover, in this study, it should be noted that during the 2-year follow-up, even though there was a deterioration in the pulmonary function parameters, this did not reach a significant difference [18].

In case of Methotrexate-induced ILD, treatment consists of drug cessation and, (especially in patients who remain symptomatic after Methotrexate withdrawal) corticosteroids. It is not advised to reinitiate Methotrexate after an ILD-event [11].

2.2. *Leflunomide*

Leflunomide acts by inhibiting the mitochondrial enzyme dihydro-orotate dehydrogenase (DHODH) via its metabolite A771726—teriflunomide [19].

Leflunomide-induced/exacerbated ILD first became a concern in 2004, when, months after the drug was released in Japan, post-marketing surveillance showed that sixteen patients developed de novo or exacerbated ILD, nine of whom died [19].

There is confounding data given the fact that Leflunomide is often a second-line therapy and most patients receiving it have a Methotrexate history. A study by Suissa et al. showed that while Leflunomide use was a risk factor for developing RA-ILD, the patients at risk had either previously used Methotrexate or had a history of ILD. Moreover, the same study noticed a prescription bias: patients with preexistent ILD were twice as likely to be prescribed Leflunomide instead of Methotrexate. Therefore, in many cases, it remains difficult to demonstrate causality when suspecting Leflunomide-induced ILD [20].

The main pathogenetic mechanism of Leflunomide-induced pneumonitis is attributed to hypersensitivity reactions [1].

Racial factors seem to have a part in Leflunomide-induced ILD: Sawada et al. reported that, in a cohort of 5054 RA patients who were prescribed LEF, 61 patients (1.2%) developed ILD as an adverse reaction to this DMARD, which was higher than that estimated at 0.02% in Western countries [21]. Similarly, Ju et al. reported a higher prevalence of Leflunomide-induced ILD: 1.0% of the Korean patients treated with Leflunomide developed this adverse reaction [22]. The frequency of leflunomide-induced ILD and its associated mortality rate are greater compared to Western countries, as indicated by this data; however, the underlying mechanisms responsible for this racial disparity in susceptibility to drug-induced ILD remains unidentified.

It has been postulated that mechanisms connected to epithelial–mesenchymal transition (EMT) might be involved in Leflunomide-induced ILD [23]. In ILD, epithelial-mesenchymal transition (EMT) might contribute to the pulmonary build-up of fibroblasts and myofibroblasts, originating from epithelial cells.

Namba's et al. research on Leflunomide's metabolite, A771726, indicated that A771726 induced EMT-like characteristics in cultured human type II alveolar (A549) cells, via DHODH inhibition. A771726 treatment resulted in the upregulation of α-smooth muscle actin (α-SMA) and Col1a1 mRNA expression, and a downregulation in the expression of E-cadherin mRNA [23].

Leflunomide-induced pneumopathy usually occurs within the first 20 weeks of therapy, being typically described as being a NSIP pattern. Risk factors associated with such lung involvement include preexisting pulmonary lesions, interstitial pneumonia, the use of a loading dose, smoking, and low body weight [21,24].

2.3. *Sulfasalazine, Hydroxychloroquineazathioprine, and Cyclosporine*

Sulfasalazine-induced lung disease is rather rare. A retrospective case study involving all patients in published reports between 1972–1999 treated with sulfasalazine and who had developed a possible pulmonary adverse reaction to the drug identified only 50 patients. In this cohort, there were only six patients with RA [25].

There is currently no available data on the pulmonary toxicity of hydroxychloroquine.

Whilst there have been case reports on pulmonary toxicity in azathioprine-treated patients [26], azathioprine has been successfully used in RA-ILD (mostly via extrapolation from the therapeutic approach in ILD secondary to the other connective tissue disorders) [27]. In a retrospective cohort analysis that studied the combined incidence rate of death, transplant and respiratory hospitalization associated with azathioprine exposure when compared with mycophenolate mofetil demonstrated a marginally better response in regard to pulmonary function tests for the azathioprine group, but with a higher rate of side effects [28]. There have been multiple case reports in which cyclosporine has been associated with sustained pulmonary function improvement as well as rheumatoid arthritis control [29–31].

3. Biologic Agents (bDMARDs)

Most bDMARDs have been linked to lung toxicity, but they have also been reported to have the potential to improve lung function and stabilize pulmonary symptoms.

Regarding biological disease-modifying antirheumatic drugs (DMARDs), tumor necrosis factor (TNF) alpha inhibitors are those with the most data on drug-induced ILD.

3.1. *Anti-TNF Drugs*

Initial concerns regarding drug-induced ILD in RA patients arose after anecdotal reports of serious exacerbations of respiratory disease following treatment with a TNF inhibitor (TNFi) in patients with pre-existing RA-ILD. Post-marketing studies have revealed that the development of ILD after TNF inhibitor therapy was a rather rare event (0.5–0.6%), but that the preexisting ILD at the moment of TNF inhibitors initiation is a risk factor for ILD exacerbations [32–34]

When analyzing a RA cohort (163 patients with RA who underwent anti-TNF therapy), Nakashita et al. demonstrated a potential risk of ILD events (progression) in patients with pre-existing ILD: 24.1% of the patients with pre-existing ILD had subsequent ILD events, whilst only 3% of the patients without pre-existing ILD registered such events (still a higher proportion than the one reported by the post-marketing studies). This data, in conjunction with descriptions regarding each of the TNF alpha inhibitors, indicates that the risk of pneumonitis seems to display a class effect [34].

In a cohort of 122 RA patients with interstitial lung disease either induced or exacerbated by TNF-targeted therapies, complete resolution was observed in up to 40% after withdrawal of the biologic agent [35].

The data available from the British Society for Rheumatology Biologics Register showed that the survival of RA-ILD patients is not influenced by anti-TNF-α therapy [36].

The imaging pattern in TNF-inhibitors-induced ILD is variable with different patterns of interstitial involvement, most commonly UIP or NSIP; cases of organizing pneumonia, diffuse alveolar damage, and lymphoid interstitial pneumonia have also been described [37].

It should also be noted that patients with RA initiated on TNF inhibitors often continue conventional DMARD therapy. In patients with concomitant Methotrexate therapy, an increased frequency of development of Methotrexate pneumonitis has been suggested [38].

Regarding the relationship between specific biological medication and drug-induced ILD, post-marketing surveillance of infliximab's safety profile in 5000 Japanese RA patients identified ILD in 25 patients (0.5%), after a mean of 2.8 infusions of infliximab, with a mean number of days from the first infusion to ILD diagnosis of 76.8 days (36–153 days) [32].

Similar data exist regarding Adalimumab: the data published by the Japan College of Rheumatology, on 3000 RA patients treated with adalimumab, reported the occurrence of interstitial pneumonia in 0.6% of patients. Additional post-marketing surveillance data on a cohort of 7440 patients placed the prevalence of ILD among adalimumab-treated patients at 0.5% [39,40].

Etanercept (ETN) was studied in a randomized controlled trial in the treatment of idiopathic pulmonary fibrosis, which showed no significant differences observed in the efficacy endpoints between the placebo and active treatment groups [41].

The post-marketing surveillance of etanercept in RA treatment, conducted by JCR (the Japan College of Rheumatology), revealed 0.6% of study patients developed ILD [33,42].

As to the impact of etanercept in combination with DMARDs, not only did the concomitant use of etanercept and methotrexate show a better RA control, but also significantly lower incidence rates for total adverse events, including ILD (when compared to etanercept as a monotherapy or associated with other DMARDs) [43]

Certolizumab, similarly to other TNFα inhibitors, may lead to acute exacerbations in RA-ILD patients, probably via NLRP3 inflammasome activation precipitate pneumonitis. The first case was reported in 2013 by Glaspole et al. [44,45] and further case reports have focused on new-onset or acute exacerbation (AE) of ILD in patients, but so far the data is limited [46].

3.2. *Alternate Mechanisms of Action (MOAs) Agents*

There is even less data regarding alternate mechanisms of action (MOAs) agents: T-cell, B-cell, and interleukin-6 inhibitors.

A study by Curtis et al., evaluating ILD incidence and exacerbation among RA patients treated with MOAs (abatacept, rituximab, and tocilizumab) compared with anti-TNFα agents, showed no significant differences regarding the risk of ILD and its related complications. It should be also noted that the patients in the MOAs group were more likely to have prior exposure to other drugs; for instance, prior biologic exposure and corticosteroid use was highest in patients in the tocilizumab and rituximab groups [47].

3.3. *Rituximab*

Rituximab seems to be a relatively safe therapy in RA patients with lung involvement. Given the fact that a worsening of preexisting ILD in RA patients treated with TNF inhibitors has been reported in the literature, Rituximab has often been the biological agent of choice in these cases. Even though data available on the matter are still rather scarce, several studies have seemed to suggest that Rituximab has a better safety profile in RA-ILD patients

In a cohort of 264 patients with RA that had received RTX, out of which 38 patients (14%) had lung involvement, lung disease remained clinically and radiologically stable in most patients, with just a single patient showing slow progression of the disease over 4 years of follow-up [48].

A cohort study involving 53 RA patients with preexisting ILD suggests that, at the very least, patients with stable ILD remain stable in regards to pulmonary function after the initiation of Rituximab therapy during prolonged follow-up [49].

Another small-scale study, which did not specifically focus on RA patients but rather on patients with ILD that did not respond to conventional immunosuppression, followed 44 patients initiated on Rituximab for a period of two years. The results of this study seem to indicate that Rituximab treatment is associated with a reduction in FVC and DLCO decline, especially when used in patients with connective-tissue-disease-associated ILD [50].

Similar results were published by Franzen et al. In a cohort of 33 RA patients treated with Rituximab, when evaluated with serial pulmonary function tests, while no instances of respiratory symptoms were reported, the DLCO showed a progressive decline during follow-up with a maximum reduction of 6.1% at 26 weeks (when compared to baseline). The risk factors for pulmonary function changes post-rituximab were cigarette smoking, repeated administration of the drug, and corticosteroid use (prednisone). The gradual decrease in DLCO suggests the possibility of subclinical pulmonary toxicity induced by rituximab, but it should be noted that both steroid use and repeated courses of rituximab are linked to a more severe RA [51].

In regards to patient mortality, the data from the British Society for Rheumatology Biologics Register for RA (BSRBR-RA) suggest that the RA-ILD who were administered rituximab as their initial biologic had superior long-term survival rates when compared to those who began treatment with TNF inhibitors, with an adjusted 5-year risk of mortality in the RTX-treated patients of approximately half that in the TNF inhibitor-treated patients, but the difference was not statistically significant (HR 0.53, 95% CI: 0.26 to 1.10) [52].

3.4. Abatacept

At the time of this review, the data on the impact of Abatacept in patients with RA-ILD are reassuring. A multicenter study including 263 RA-ILD patients treated with Abatacept showed an improvement or stabilization after 1 year of therapy regarding dyspnea scores (in 91.9%), FVC (in 87.7%), DLCO (in 90.6%), and chest HRCT (in 76.6%) [53].

A prospective study on a cohort of 57 RA-ILD patients treated with abatacept for a median (IQR) of 27.3 (12.2–42.8) months demonstrated both arthritis control and pulmonary involvement stabilization in 71% of patients. The factors associated with lung disease progression and mortality were high disease activity (calculated with the DAS28-ESR formula) and low baseline DLCO and FVC values [54].

In another cohort of 44 patients with RA-ILD treated with Abatacept, only 11.4% experienced a significant worsening during the 18-month follow-up period (when evaluated on the change in the percentage of fibrosis). The variables associated with RA-ILD worsening were current smoker status and concomitant Methotrexate use [55].

Another study that compared the effect of biological therapies on airway and interstitial lung disease in RA patients found abatacept to be an independent protective factor for both RA-ILD, but also for airway disease exacerbation [56].

When comparing the safety of Abatacept in monotherapy with ABA associated with synthetic DMARDs in RA patients with interstitial lung disease, both strategies seem to have similar safety and effectiveness profiles. In a retrospective multicenter study of RA-ILD Caucasian patients treated with abatacept either as monotherapy associated with Methotrexate or non-Methotrexate DMARDs, more than 70% (out of 263) of the patients had stable or improved ILD after 18 months of treatment, with no significant differences between treatment groups [57].

3.5. Tocilizumab

There are data suggesting that uncontrolled arthritis activity during tocilizumab treatment could lead to an acute exacerbation of RA-ILD. A retrospective, case-controlled study involved 395 consecutive RA patients treated with tocilizumab, 78 with ILD, and 317 without ILD. Six acute exacerbations occurred, after a median treatment duration of

48 weeks. The patients experiencing acute ILD exacerbations had higher disease activity measured using the Clinical Disease Activity Index (CDAI) at 24 weeks (20.8 vs. 6.2, $p = 0.019$), suggesting that the acute exacerbations of RA-ILD were likely attributable to uncontrolled disease activity rather than an adverse effect of the drug [58]

In the post-marketing data of a cohort of 7901 RA patients, with a cumulative exposure to tocilizumab of 3831.8 patient-years (PY), 38 (0.5%, with an incidence rate of 1.0 event per 100 person-years) were reported to have interstitial lung disease (ILD), with 22 of them having either a concurrent or prior medical history of ILD at the beginning of the study. It should be noted also that 24 of the ILD patients had a prior history of biological DMARDs. Multivariate logistic regression analysis identified advanced age (≥65 years) and previous or concurrent ILD at baseline as the risk factors for ILD [59].

In a study that included 125 elderly (>65 years) patients with RA treated either with Abatacept (n = 47) or Tocilizumab (n = 78), the most common adverse event that resulted in the discontinuation of Tocilizumab treatment was ILD. Out of the five patients who developed ILD while on Tocilizumab, two had a worsening of pre-existing ILD, while four had been concurrently using Methotrexate. In contrast, ILD was not reported as an adverse event leading to the discontinuation of abatacept in elderly patients with RA [60].

In a cohort of 11,219 patients, totaling 13,795 episodes of biologic exposure, there were no significant differences in the risk of ILD and its related complications between RA patients receiving different classes of biological therapy. The data suggests that the incidence and exacerbation of ILD in the Tocilizumab and Rituximab groups may be exaggerated by treatment resistance or by the severity of the disease, since the first line of biologics were anti-TNF agents in most patients [47].

3.6. Anakinra

Information regarding anakinra's pulmonary effects in RA patients is scarce, and so far, there have not been any randomized controlled trials in regard to RA-ILD. The data scarcity may be because anakinra use has seen a significant decline over time. The newer therapeutic options seem to provide better RA control, a better safety profile, and more convenient dosing options.

4. Targeted-Synthetic DMARDS/JANUS KINASE Inhibitors

Regarding targeted-synthetic DMARDs, which represent a somewhat newer alternative in RA treatment, information pertaining to interstitial lung disease and pulmonary safety is currently limited, but the results published thus far seem to indicate a low rate of ILD development during Janus-kinase inhibitor use [61–63].

Moreover, there is data regarding Tofacitinib: a post hoc analysis of 21 trials, including 7061 patients (patient-years of exposure 23,393.7) who received tofacitinib, showed that the IR for an ILD event was 0.18 for both tofacitinib 5 mg BID and 10 mg BID, an incidence associated with known risk factors of RA-ILD, such as age, smoker status, and high disease activity (DAS28 scores) [62].

A retrospective study using claims data pertaining to 28,559 patients with RA from the Optum Clinformatics Data Mart Database showed that, when calculating the IRs per 1000 person-years for ILD, patients who received tofacitinib were 69% less likely to develop ILD (IR = 1.48) compared to those treated with adalimumab (IR = 4.30) [63].

In a retrospective study of 75 patients with RA and ILD treated either with Janus kinase inhibitors (tofacitinib 5 mg BID or baricitinib 4 mg daily) or abatacept (125 mg/week) for at least 18 months, the data showed similar safety profiles in regard to pulmonary outcomes. RA-ILD stability or improvement was reached in 83.9% and 88.6% of patients, respectively. In the multivariate regression analysis, the only variable related to RA-ILD deterioration in patients treated with Janus kinase inhibitors was disease duration ($p < 0.001$) [64].

In a descriptive, multicentric, retrospective cohort study that included data pooled from eight randomized trials, with 3770 RA patients treated with baricitinib, with 12,358 patient-years of exposure, Salvarani et al. identified 21 ILD cases with an exposure-adjusted incidence rate (EAIR) of 0.17 per 100 patient-years of exposure (PYE), proving a low risk of developing ILD during treatment [65].

Furthermore, in a retrospective exploratory study that analyzed a cohort of 43 RA-ILD patients treated with either baricitinib (65.12%), tofacitinib (20.93%), filgotinib (6.98%), or upadacitinib (6.98%), the DLCO improved or remained stable in 80% of the cases. The forced vital capacity followed the same trend, as it worsened in only 10.71% of the patients and the chest HRCT demonstrated a progression of ILD in 9.30%. While combined therapy with methotrexate was documented in 38.10% of the patients, no improvement or deterioration was seen in the HRCT and pulmonary function tests between those with monotherapy or those with concomitant methotrexate use [66].

These results suggest that treatment with Janus kinase inhibitors may provide benefit in reducing the risk of developing RA-ILD.

5. Conclusions and Research Agenda

The establishment of treatment guidelines for RA-ILD continues to present an ongoing challenge, with further complications arising from the potential pulmonary toxicity of both synthetic and biologic DMARDs. Our review will hopefully be able to aid the clinician in choosing the optimal treatment strategy for both at-risk patients and those with established ILD.

Both initial and subsequent evaluations of RA patients should consider both articular and extraarticular manifestations. Risk factors (such as smoking or environmental exposures), clinical findings, dyspnea scales, the 6-min walk test, pulmonary function tests (PFT), DLCO, and imaging findings should be documented in order to assess the severity of pulmonary involvement and to detect its progression.

Routine clinical evaluations of RA patients should include pulmonary evaluation, with an emphasis on a structured follow-up plan for individuals with identified ILD or those at high risk, including regular clinical evaluations, PFTs, and imaging assessment.

Optimizing therapy and follow-up strategies require a multidisciplinary approach that involves, at a minimum, a rheumatologist, a pulmonologist, and a radiologist. Drug-induced lung damage should also be considered in any RA patient who develops respiratory symptoms or new imaging changes during therapy.

In patients with low disease activity/remission and with stable pulmonary involvement, the ongoing therapy should be continued; however, in patients with progressive ILD and active joint disease, switching to JAK inhibitors or alternate mechanisms of action (MOAs) agents could be beneficial.

During the past two decades, we have witnessed many exciting developments in the treatment of rheumatoid arthritis. While this is very encouraging, ongoing research regarding the long-term safety profile of these agents is essential, especially among patients with extra-articular manifestations such as RA-ILD, or patients with preexisting pulmonary conditions. Moreover, there is a need for clinical registries/randomized controlled trials on antifibrotic agents and DMARD associations.

Author Contributions: All authors have contributed equally to the article. All authors have read and agreed to the published version of the manuscript.

Funding: This research received no external funding.

Institutional Review Board Statement: Not applicable.

Informed Consent Statement: Not applicable.

Data Availability Statement: Not applicable.

Acknowledgments: There is no funding or administrative support provided to the authors.

Conflicts of Interest: All the authors declare no conflict of interest.

References

1. Nakamura, H.; Minoru, K. Definition and Pathogenesis of Drug-Induced Lung Injury: What Is DLI? In *Drug-Induced Lung Injury*; Hanaoka, M., Nakamura, H., Aoshiba, K., Eds.; Respiratory Disease Series: Diagnostic Tools and Disease Managements; Springer: Singapore, 2018; pp. 3–12. [CrossRef]
2. Dias, O.M.; Pereira, D.A.S.; Baldi, B.G.; Costa, A.N.; Athanazio, R.A.; Kairalla, R.A.; Carvalho, C.R.R. Adalimumab-Induced Acute Interstitial Lung Disease in a Patient with Rheumatoid Arthritis. *J. Bras. Pneumol.* **2014**, *40*, 77–81. [CrossRef] [PubMed]
3. Kubo, K.; Azuma, A.; Kanazawa, M.; Kameda, H.; Kusumoto, M.; Genma, A.; Saijo, Y.; Sakai, F.; Sugiyama, Y.; Tatsumi, K.; et al. Consensus statement for the diagnosis and treatment of drug-induced lung injuries. *Respir. Investig.* **2013**, *51*, 260–277. [CrossRef] [PubMed]
4. Matsuno, O. Drug-induced interstitial lung disease: Mechanisms and best diagnostic approaches. *Respir. Res.* **2012**, *13*, 39. [CrossRef] [PubMed]
5. Pneumotox. Drug. Available online: https://www.pneumotox.com/drug/index/ (accessed on 10 April 2023).
6. Proudman, S.; Lake, F. Rheumatoid Arthritis and Lung Disease: From Mechanisms to a Practical Approach. *Semin. Respir. Crit. Care Med.* **2014**, *35*, 222–238. [CrossRef]
7. Smolen, J.S.; Aletaha, D.; Bijlsma, J.W.J.; Breedveld, F.C.; Boumpas, D.; Burmester, G.; Combe, B.; Cutolo, M.; de Wit, M.; Dougados, M.; et al. Treating rheumatoid arthritis to target: Recommendations of an international task force. *Ann. Rheum. Dis.* **2010**, *69*, 631–637. [CrossRef]
8. Handa, R. Biologics in Rheumatoid Arthritis. In *Handbook of Biologics for Rheumatological Disorders*; Springer Nature: Singapore, 2022; pp. 13–21.
9. Conway, R.; Low, C.; Coughlan, R.J.; O'Donnell, M.J.; Carey, J.J. Methotrexate and lung disease in rheumatoid arthritis: A meta-analysis of randomized controlled trials. *Arthritis Rheumatol.* **2014**, *66*, 803–812. [CrossRef]
10. Handa, T.; Yonezawa, A.; Azuma, A. Epidemiology and Risk Factors of Drug-Induced Lung Disease: What Are the Prevalence and Risk Factors of DILD? In *Drug-Induced Lung Injury*; Springer: Singapore, 2017; pp. 13–26. [CrossRef]
11. Roubille, C.; Boulos, H. Interstitial Lung Diseases Induced or Exacerbated by DMARDS and Biologic Agents in Rheumatoid Arthritis: A Systematic Literature Review. *Semin. Arthritis Rheum.* **2014**, *43*, 613–626. [CrossRef]
12. Salliot, C.; van der Heijde, D. Long-term safety of methotrexate monotherapy in patients with rheumatoid arthritis: A systematic literature research. *Ann. Rheum. Dis.* **2008**, *68*, 1100–1104. [CrossRef]
13. Chikura, B.; Sathi, N.; Lane, S.; Dawson, J.K. Variation of immunological response in methotrexate-induced pneumonitis. *Rheumatology* **2008**, *47*, 1647–1650. [CrossRef]
14. Sathi, N.; Chikura, B.; Kaushik, V.V.; Wiswell, R.; Dawson, J.K. How common is methotrexate pneumonitis? A large prospective study investigates. *Clin. Rheumatol.* **2012**, *31*, 79–83. [CrossRef]
15. Kameda, H. DLI Caused by Disease-Modifying Antirheumatic Drugs: What Are the Characteristics of DLI by Disease-Modifying Antirheumatic Drugs? In *Drug-Induced Lung Injury*; Springer: Singapore, 2018; pp. 165–176.
16. Searles, G.; McKendry, R.J. Methotrexate pneumonitis in rheumatoid arthritis: Potential risk factors. Four case reports and a review of the literature. *J. Rheumatol.* **1987**, *14*, 1164–1171. [PubMed]
17. Lateef, O.; Shakoor, N.; Balk, A.R. Methotrexate pulmonary toxicity. *Expert Opin. Drug Saf.* **2005**, *4*, 723–730. [CrossRef] [PubMed]
18. Dawson, J.K.; Graham, D.R.; Desmond, J.; Fewins, H.E.; Lynch, M.P. Investigation of the chronic pulmonary effects of low-dose oral methotrexate in patients with rheumatoid arthritis: A prospective study incorporating HRCT scanning and pulmonary function tests. *Rheumatology* **2002**, *41*, 262–267. [CrossRef] [PubMed]
19. McCurry, J. Japan deaths spark concerns over arthritis drug. *Lancet* **2004**, *363*, 461. [CrossRef] [PubMed]
20. Suissa, S.; Hudson, M.; Ernst, P. Faculty Opinions recommendation of Leflunomide use and the risk of interstitial lung disease in rheumatoid arthritis. *Arthritis Rheumatol.* **2006**, *54*, 1435–1439. [CrossRef] [PubMed]
21. Sawada, T.; Inokuma, S.; Sato, T.; Otsuka, T.; Saeki, Y.; Takeuchi, T.; Matsuda, T.; Takemura, T.; Sagawa, A. Study Committee for Leflunomide-induced Lung Injury, Japan College of Rheumatology. Leflunomide-induced interstitial lung disease: Prevalence and risk factors in Japanese patients with rheumatoid arthritis. *Rheumatology* **2009**, *48*, 1069–1072. [CrossRef]
22. Ju, J.H.; Kim, S.-I.; Lee, J.-H.; Lee, S.-I.; Yoo, W.-H.; Choe, J.-Y.; Chung, S.-H.; Lee, J.; Lee, Y.-H.; Lee, S.-S.; et al. Risk of interstitial lung disease associated with leflunomide treatment in Korean patients with rheumatoid arthritis. *Arthritis Rheum.* **2007**, *56*, 2094–2096. [CrossRef]
23. Namba, T.; Tanaka, K.-I.; Ito, Y.; Hoshino, T.; Matoyama, M.; Yamakawa, N.; Isohama, Y.; Azuma, A.; Mizushima, T. Induction of EMT-like phenotypes by an active metabolite of leflunomide and its contribution to pulmonary fibrosis. *Cell Death Differ.* **2010**, *17*, 1882–1895. [CrossRef] [PubMed]
24. Chikura, B.; Lane, S.; Dawson, J.K. Clinical expression of leflunomide-induced pneumonitis. *Rheumatology* **2009**, *48*, 1065–1068. [CrossRef]
25. Parry, S.D.; Barbatzas, C.; Peel, E.T.; Barton, J.R. Sulphasalazine and lung toxicity. *Eur. Respir. J.* **2002**, *19*, 756–764. [CrossRef]
26. Ishida, T.; Kotani, T.; Takeuchi, T.; Makino, S. Pulmonary toxicity after initiation of azathioprine for treatment of interstitial pneumonia in a patient with rheumatoid arthritis. *J. Rheumatol.* **2012**, *39*, 1104–1105. [CrossRef]
27. Boerner, E.B.; Cuyas, M.; Theegarten, D.; Ohshimo, S.; Costabel, U.; Bonella, F. Azathioprine for Connective Tissue Disease-Associated Interstitial Lung Disease. *Respiration* **2020**, *99*, 628–636. [CrossRef]

28. Oldham, J.M.; Lee, C.; Valenzi, E.; Witt, L.J.; Adegunsoye, A.; Hsu, S.; Chen, L.; Montner, S.; Chung, J.H.; Noth, I.; et al. Azathioprine response in patients with fibrotic connective tissue disease-associated interstitial lung disease. *Respir. Med.* **2016**, *121*, 117–122. [CrossRef]
29. Suda, T. Up-to-Date Information on Rheumatoid Arthritis-Associated Interstitial Lung Disease. *Clin. Med. Insights Circ. Respir. Pulm. Med.* **2015**, *9s1*, 155–162. [CrossRef]
30. Cassone, G.; Manfredi, A.; Vacchi, C.; Luppi, F.; Coppi, F.; Salvarani, C.; Sebastiani, M. Treatment of Rheumatoid Arthritis-Associated Interstitial Lung Disease: Lights and Shadows. *J. Clin. Med.* **2020**, *9*, 1082. [CrossRef]
31. Chang, H.K.; Park, W.; Ryu, D.S. Successful Treatment of Progressive Rheumatoid Interstitial Lung Disease with Cyclosporine: A Case Report. *J. Korean Med. Sci.* **2002**, *17*, 270–273. [CrossRef] [PubMed]
32. Takeuchi, T.; Tatsuki, Y.; Nogami, Y.; Ishiguro, N.; Tanaka, Y.; Yamanaka, H.; Kamatani, N.; Harigai, M.; Ryu, J.; Inoue, K.; et al. Postmarketing surveillance of the safety profile of infliximab in 5000 Japanese patients with rheumatoid arthritis. *Ann. Rheum. Dis.* **2008**, *67*, 189–194. [CrossRef] [PubMed]
33. Koike, T.; Harigai, M.; Inokuma, S.; Ishiguro, N.; Ryu, J.; Takeuchi, T.; Tanaka, Y.; Yamanaka, H.; Fujii, K.; Yoshinaga, T.; et al. Postmarketing surveillance of safety and effectiveness of etanercept in Japanese patients with rheumatoid arthritis. *Mod. Rheumatol.* **2011**, *21*, 343–351. [CrossRef]
34. Nakashita, T.; Ando, K.; Kaneko, N.; Takahashi, K.; Motojima, S. Potential risk of TNF inhibitors on the progression of interstitial lung disease in patients with rheumatoid arthritis. *BMJ Open* **2014**, *4*, e005615. [CrossRef] [PubMed]
35. Perez-Alvarez, R.; Perez-de-Lis, M.; Diaz-Lagares, C.; Pego-Reigosa, J.M.; Retamozo, S.; Bove, A.; Brito-Zeron, P.; Bosch, X.; Ramos-Casals, M. Interstitial lung disease induced or exacerbated by TNF-targeted therapies: Analysis of 122 cases. *Semin. Arthritis Rheum.* **2011**, *41*, 256–264. [CrossRef]
36. Dixon, W.G.; Hyrich, K.L.; Watson, K.D.; Lunt, M.; Symmons, D.P.M.; BSRBR Control Centre Consortium. Influence of anti-TNF therapy on mortality in patients with rheumatoid arthritis-associated interstitial lung disease: Results from the British Society for Rheumatology Biologics Register. *Ann. Rheum. Dis.* **2010**, *69*, 1086–1091. [CrossRef] [PubMed]
37. Cavagna, L.; Monti, S.; Grosso, V.; Boffini, N.; Scorletti, E.; Crepaldi, G.; Caporali, R. The Multifaceted Aspects of Interstitial Lung Disease in Rheumatoid Arthritis. *BioMed. Res. Int.* **2013**, *2013*, 759760. [CrossRef] [PubMed]
38. Khasnis, A.A.; Calabrese, L.H. Tumor Necrosis Factor Inhibitors and Lung Disease: A Paradox of Efficacy and Risk. *Semin. Arthritis Rheum.* **2010**, *40*, 147–163. [CrossRef]
39. Koike, T.; Harigai, M.; Ishiguro, N.; Inokuma, S.; Takei, S.; Takeuchi, T.; Yamanaka, H.; Tanaka, Y. Safety and effectiveness of adalimumab in Japanese rheumatoid arthritis patients: Postmarketing surveillance report of the first 3000 patients. *Mod. Rheumatol.* **2012**, *22*, 498–508. [CrossRef] [PubMed]
40. Koike, T.; Harigai, M.; Ishiguro, N.; Inokuma, S.; Takei, S.; Takeuchi, T.; Yamanaka, H.; Haruna, S.; Ushida, N.; Kawana, K.; et al. Safety and effectiveness of adalimumab in Japanese rheumatoid arthritis patients: Postmarketing surveillance report of 7740 patients. *Mod. Rheumatol.* **2014**, *24*, 390–398. [CrossRef]
41. Raghu, G.; Brown, K.K.; Costabel, U.; Cottin, V.; Du Bois, R.M.; Lasky, J.A.; Thomeer, M.; Utz, J.P.; Khandker, R.K.; McDermott, L.; et al. Faculty Opinions recommendation of Treatment of idiopathic pulmonary fibrosis with etanercept: An exploratory, placebo-controlled trial. *Am. J. Respir. Crit. Care Med.* **2008**, *178*, 948–955. [CrossRef]
42. Horai, Y.; Miyamura, T.; Shimada, K.; Takahama, S.; Minami, R.; Yamamoto, M.; Suematsu, E. Eternacept for the treatment of patients with rheumatoid arthritis and concurrent interstitial lung disease. *J. Clin. Pharm. Ther.* **2012**, *37*, 117–121. [CrossRef]
43. Koike, T.; Harigai, M.; Inokuma, S.; Ishiguro, N.; Ryu, J.; Takeuchi, T.; Tanaka, Y.; Yamanaka, H.; Hirose, T.; Yoshinaga, T.; et al. Safety and Effectiveness of 6 Months' Etanercept Monotherapy and Combination Therapy in Japanese Patients with Rheumatoid Arthritis: Effect of Concomitant Disease-modifying Antirheumatic Drugs. *J. Rheumatol.* **2013**, *40*, 1658–1668. [CrossRef]
44. Migita, K.; Tsuji, Y.; Hisatomi, K.; Shigeno, R.; Izumi, Y.; Iwanaga, N.; Koga, T. Acute exacerbation of rheumatoid interstitial lung disease during the maintenance therapy with certolizumab pegol. *Mod. Rheumatol.* **2017**, *27*, 1079–1082. [CrossRef]
45. Glaspole, I.N.; Hoy, R.F.; Ryan, P.F. A case of certolizumab-induced interstitial lung disease in a patient with rheumatoid arthritis. *Rheumatology* **2013**, *52*, 2302–2304. [CrossRef]
46. Savage, E.M.; Millar, A.M.; Taggart, A.J. Comment on: A case of certolizumab-induced interstitial lung disease in a patient with rheumatoid arthritis. *Rheumatology* **2014**, *53*, 1154–1155. [CrossRef] [PubMed]
47. Curtis, J.R.; Sarsour, K.; Napalkov, P.; Costa, L.A.; Schulman, K.L. Incidence and complications of interstitial lung disease in users of tocilizumab, rituximab, abatacept and anti-tumor necrosis factor α agents, a retrospective cohort study. *Arthritis Res. Ther.* **2015**, *17*, 319. [CrossRef] [PubMed]
48. Becerra, E.; Cambridge, G.; Leandro, M. FRI0228 Safety and efficacy of rituximab in patients with rheumatoid arthritis and lung involvement. *Ann. Rheum. Dis.* **2013**, *72*, 220. [CrossRef]
49. Yusof, M.; Kabia, A.; Dass, S.; Vital, E.; Beirne, P.; Emery, P. THU0158 Efficacy and Safety of Rituximab in Rheumatoid Arthritis Patients with Concomitant Interstitial Lung Disease: 10-Year Experience at Single Centre. *Ann. Rheum. Dis.* **2015**, *74* (Suppl. S2), 251. [CrossRef]
50. Kokosi, M.; Keir, G.J.; Corte, T.J.; Troy, L.; Saunders, P.; Chua, F.; Maher, T.M.; Renzoni, E.A.; Wells, A.U. Rituximab in severe, progressive interstitial lung disease. *Eur. Respir. J.* **2016**, *48* (Suppl. 60), PA4886. [CrossRef]
51. Franzen, D.; Ciurea, A.; Bratton, D.J.; Clarenbach, C.F.; Latshang, T.D.; Russi, E.W.; Kyburz, D.; Kohler, M. Effect of rituximab on pulmonary function in patients with rheumatoid arthritis. *Pulm. Pharmacol. Ther.* **2016**, *37*, 24–29. [CrossRef]

52. Druce, K.L.; Iqbal, K.; Watson, K.D.; Symmons, D.P.M.; Hyrich, K.L.; Kelly, C. Mortality in patients with interstitial lung disease treated with rituximab or TNFi as a first biologic. *RMD Open* **2017**, *3*, e000473. [CrossRef]
53. Fernández-Díaz, C.; Castañeda, S.; Melero-González, R.B.; Ortiz-Sanjuán, F.; Juan-Mas, A.; Carrasco-Cubero, C.; Casafont-Solé, I.; Olivé, A.; Rodríguez-Muguruza, S.; Almodóvar-González, R.; et al. Abatacept in interstitial lung disease associated with rheumatoid arthritis: National multicenter study of 263 patients. *Rheumatology* **2020**, *59*, 3906–3916. [CrossRef]
54. Mena-Vázquez, N.; Rojas-Gimenez, M.; Fuego-Varela, C.; García-Studer, A.; Perez-Gómez, N.; Romero-Barco, C.M.; Godoy-Navarrete, F.J.; Manrique-Arija, S.; Gandía-Martínez, M.; Calvo-Gutiérrez, J.; et al. Safety and Effectiveness of Abatacept in a Prospective Cohort of Patients with Rheumatoid Arthritis–Associated Interstitial Lung Disease. *Biomedicines* **2022**, *10*, 1480. [CrossRef]
55. Tardella, M.; Di Carlo, M.; Carotti, M.; Giovagnoni, A.; Salaffi, F. Abatacept in rheumatoid arthritis-associated interstitial lung disease: Short-term outcomes and predictors of progression. *Clin. Rheumatol.* **2021**, *40*, 4861–4867. [CrossRef]
56. Kurata, I.; Tsuboi, H.; Terasaki, M.; Shimizu, M.; Toko, H.; Honda, F.; Ohyama, A.; Yagishita, M.; Osada, A.; Ebe, H.; et al. Effect of biological disease-modifying anti-rheumatic drugs on airway and interstitial lung disease in patients with rheumatoid arthritis. *Intern. Med.* **2019**, *58*, 1703–1712. [CrossRef]
57. Fernández-Díaz, C.; Atienza-Mateo, B.; Castañeda, S.; Melero-Gonzalez, R.B.; Ortiz-SanJuan, F.; Loricera, J.; Casafont-Solé, I.; Rodríguez-García, S.; Aguilera-Cros, C.; Villa-Blanco, I.; et al. Abatacept in monotherapy vs combined in interstitial lung disease of rheumatoid arthritis—Multicentre study of 263 Caucasian patients. *Rheumatology* **2022**, *61*, 299–308. [CrossRef]
58. Akiyama, M.; Kaneko, Y.; Yamaoka, K.; Kondo, H.; Takeuchi, T. Association of disease activity with acute exacerbation of interstitial lung disease during tocilizumab treatment in patients with rheumatoid arthritis: A retrospective, case–control study. *Rheumatol. Int.* **2016**, *36*, 881–889. [CrossRef]
59. Koike, T.; Harigai, M.; Inokuma, S.; Ishiguro, N.; Ryu, J.; Takeuchi, T.; Takei, S.; Tanaka, Y.; Sano, Y.; Yaguramaki, H.; et al. Effectiveness and safety of tocilizumab: Postmarketing surveillance of 7901 patients with rheumatoid arthritis in Japan. *J. Rheumatol.* **2014**, *41*, 15–23. [CrossRef]
60. Temmoku, J.; Miyata, M.; Suzuki, E.; Sumichika, Y.; Saito, K.; Yoshida, S.; Matsumoto, H.; Fujita, Y.; Matsuoka, N.; Asano, T.; et al. Comparing the effectiveness and safety of Abatacept and Tocilizumab in elderly patients with rheumatoid arthritis. *PLoS ONE* **2022**, *17*, e0274775. [CrossRef]
61. Citera, G.; Mysler, E.; Madariaga, H.; Cardiel, M.H.; Castañeda, O.; Fischer, A.; Richette, P.; Chartrand, S.; Park, J.K.; Strengholt, S.; et al. Low Interstitial Lung Disease Event Rate in Patients with Rheumatoid Arthritis: Pooled Post Hoc Analysis of Data from the Tofacitinib Clinical Development Program [abstract]. *Arthritis Rheumatol.* **2018**, *70* (Suppl. S10), 525.
62. Citera, G.; Mysler, E.; Madariaga, H.; Cardiel, M.H.; Castañeda, O.; Fischer, A.; Richette, P.; Chartrand, S.; Park, J.K.; Strengholt, S.; et al. Incidence rates of interstitial lung disease events in tofacitinib-treated rheumatoid arthritis patients: Post hoc analysis from 21 clinical trials. *J. Clin. Rheumatol.* **2021**, *27*, e482. [CrossRef] [PubMed]
63. Baker, M.C.; Liu, Y.; Lu, R.; Lin, J.; Melehani, J.; Robinson, W.H. Incidence of Interstitial Lung Disease in Patients with Rheumatoid Arthritis Treated With Biologic and Targeted Synthetic Disease-Modifying Antirheumatic Drugs. *JAMA Netw. Open* **2023**, *6*, e233640. [CrossRef] [PubMed]
64. Tardella, M.; Di Carlo, M.; Carotti, M.; Ceccarelli, L.; Giovagnoni, A.; Salaffi, F. A retrospective study of the efficacy of JAK inhibitors or abatacept on rheumatoid arthritis-interstitial lung disease. *Inflammopharmacology* **2022**, *30*, 705–712. [CrossRef] [PubMed]
65. Salvarani, C.; Sebastiani, M.; Dieude, P.; Garcia, M.; Deberdt, W.; Rogai, V.; de la Torre, I.; Inciarte-Mundo, J.; Balsa, A. Baricitinib and the risk of incident interstitial lung disease: A descriptive clinical case report from clinical trials. *Rheumatol. Ther.* **2021**, *8*, 1435–1441. [CrossRef]
66. Venerito, V.; Manfredi, A.; Carletto, A.; Gentileschi, S.; Atzeni, F.; Guiducci, S.; Lavista, M.; La Corte, L.; Pedrollo, E.; Scardapane, A.; et al. Evolution of Rheumatoid-Arthritis-Associated Interstitial Lung Disease in Patients Treated with JAK Inhibitors: A Retrospective Exploratory Study. *J. Clin. Med.* **2023**, *12*, 957. [CrossRef] [PubMed]

biomedicines

MDPI

Review

Clinicopathological Outlines of Post-COVID-19 Pulmonary Fibrosis Compared with Idiopathic Pulmonary Fibrosis

Roxana-Elena Cîrjaliu [1,2], Mariana Deacu [2,3,*], Ioana Gherghișan [1,4,*], Angela-Ștefania Marghescu [5,6], Manuela Enciu [2,3], Gabriela Izabela Băltățescu [7], Antonela Anca Nicolau [7], Doina-Ecaterina Tofolean [1,2], Oana Cristina Arghir [1,4] and Ariadna-Petronela Fildan [1,4]

1 Department of Pneumology, Faculty of Medicine, "Ovidius" University of Constanta, 900470 Constanta, Romania; roxana.cirjaliu@gmail.com (R.-E.C.); tofoleandoina@yahoo.com (D.-E.T.); arghir_oana@yahoo.com (O.C.A.); ariadnapetrofildan@yahoo.com (A.-P.F.)
2 Clinical Emergency "St. Andrew" Hospital of Constanta, 900591 Constanta, Romania; iftimemanuela@yahoo.com
3 Department of Anatomopathology, Faculty of Medicine, "Ovidius" University of Constanta, 900470 Constanta, Romania
4 Pneumology Hospital of Constanta, 900002 Constanta, Romania
5 Department of Anatomopathology, "Carol Davila" University of Medicine and Pharmacy, 020021 Bucharest, Romania; angela.varban@drd.umfcd.ro
6 Pneumology Institute "Marius Nasta", 50158 Bucharest, Romania
7 Center for Research and Development of the Morphological and Genetic Studies of Malignant Pathology-CEDMOG, "Ovidius" University of Constanta, 900591 Constanta, Romania; gabrielabaltatescu@yahoo.com (G.I.B.); ancanicolau@rocketmail.com (A.A.N.)
* Correspondence: deacu_mariana@yahoo.com (M.D.); ioana.husaru@gmail.com (I.G.)

Abstract: This review brings together the current knowledge regarding the risk factors and the clinical, radiologic, and histological features of both post-COVID-19 pulmonary fibrosis (PCPF) and idiopathic pulmonary fibrosis (IPF), describing the similarities and the disparities between these two diseases, using numerous databases to identify relevant articles published in English through October 2022. This review would help clinicians, pathologists, and researchers make an accurate diagnosis, which can help identify the group of patients selected for anti-fibrotic therapies and future therapeutic perspectives.

Keywords: post-COVID-19 pulmonary fibrosis; idiopathic pulmonary fibrosis; interstitial lung disease; pulmonary fibrosis

1. Introduction

The novel "severe acute respiratory syndrome coronavirus 2" (SARS-CoV-2) is the cause of the coronavirus disease (COVID-19), which has become a global pandemic causing millions of deaths. SARS-CoV-2 can affect various organs in the body, leading to acute organ damage and long-term sequelae [1,2]. Three years after the start of the pandemic, prospective studies regarding the long-term changes of SARS-CoV-2 infection have just begun to emerge. While clinical studies regarding the safety and effectiveness of antiviral agents and vaccines are ongoing, chronic pulmonary consequences of COVID-19 are increasingly concerning and have become more recognized.

Although pulmonary fibrosis has been observed in varying degrees of evolution in patients with SARS-CoV-2 infection, its mechanism has not been sufficiently studied and elucidated, with various causes being proposed, such as chronic inflammation and idiopathic, genetic, and age-related fibroproliferative processes. The data showed that 40% of patients with COVID-19 develop ARDS, while only 20% have severe outcomes. Pulmonary fibrosis can be a sequel of ARDS, although the radiological anomalies could be of little clinical significance and diminished after lung ventilation [3]. The prevalence of post-COVID-19 fibrosis will be visible in time, but early analysis suggests that more than a third of the recovered patients develop fibrotic anomalies. One study showed that 4% of patients with a disease duration of more than a week, 24% with a disease duration between 1 and 3 weeks, and 61% with a disease duration longer than 3 weeks develop fibrosis [4].

Retrospective studies have shown that over 90% of hospitalized patients have presented persistent modifications of the lungs at the moment of discharge from the hospital [5,6]. In contrast, most patients with mild forms of COVID-19 had ground glass opacities that disappeared three weeks after hospital discharge [7]. In a series of severe COVID-19 patients who had recovered, at the one-year follow-up, nearly 25% had persistent radiologic abnormalities with features characteristic of pulmonary fibrosis, including reticular opacities, septal thickening, and traction bronchiectasis [5]. Although pulmonary fibrosis is not a common complication in other forms of viral pneumonia, it might be a possible complication of COVID-19 pneumonia leading to irreversible lung dysfunction in many COVID-19 patients [8].

Pulmonary fibrosis occurs when the tissue's restoration ability is affected by the impaired regeneration of the alveolar epithelium, with fibroblast persistence and excessive collagen deposition, leading to abnormal lung architecture [9]. Different potential contributing risk factors for persistent fibrotic lung changes in COVID-19 patients were proposed, such as age over 40 years old, hospitalization duration over 16 days, noninvasive mechanical ventilation, tachycardia at admission, acute respiratory distress syndrome (ARDS), and computer tomography (CT) score over 17 at the initial evaluation [10]. Due to the tremendous number of COVID-19 cases and the disease severity, it is crucial to consider the potential long-term implications of this disease.

Interstitial lung diseases (ILDs) are a heterogeneous group of parenchymal lung diseases characterized by varying degrees of inflammation and fibrosis. Idiopathic pulmonary fibrosis (IPF) having an unknown origin, a chronic progressive evolution, and a poor prognosis, with a mean survival of 2.5–5 years after definite diagnosis, represents around 17–37% of diffuse interstitial lung diseases [11]. It primarily affects older adults and has exclusively pulmonary involvement, causing chronic, progressive lung scarring, as defined by the pathological histopathologic and/or radiologic pattern of usual interstitial pneumonia (UIP). The histological characteristics of the UIP pattern are marked parenchymal fibrosis with or without "honeycombing", predominantly subpleural and paraseptal, the presence of fibroblast foci, traction bronchiectasis, and remodeling of the alveoli [12].

Regarding etiopathogenesis, current theories hypothesize that alveolar epithelial cell injury is the initiating factor and, in this sense, is essential for excluding another interstitial lung disease. The lesion consists of excessive proliferation of fibroblasts and myofibroblasts and deposition of disorganized collagen and extracellular matrix, resulting in lung architecture distortion, with or without honeycomb cyst formation [13].

In conclusion, PCPF and IPF determine lung fibrosis through different mechanisms, characterized by altered tissue regeneration, causing abnormal lung parenchyma. There is a need for further studies to establish the features of post-COVID-19 pulmonary fibrosis (PCPF) and its evolution, such as the permanent or progressive character that appears in other fibrotic lung diseases such as idiopathic pulmonary fibrosis (IPF).

2. Materials and Methods

The scope of our study is to make an accurate comparison between PCPF and IPF from etiopathological, clinical, and paraclinical points of view, to analyze the therapeutic interventions that can be used for PCPF. For our study, the eligibility criteria included the inclusion and exclusion criteria, which referred to the type of study, databases, date, language, and search terms. We selected the following studies: systematic reviews, meta-analyses, systematic literature reviews, systematic narrative reviews, systematic scoping reviews, systematic meta-review, systematic evidence reviews, systematic critical reviews, systematic integrative reviews, case reports, and case series.

We performed this literature review using the following databases: MEDLINE, Clarivate, PubMed, Scopus, Google Scholar, and Science Direct, to identify relevant articles published in English through October 2022. We used Medical Subject Headings (MeSH), standardized biomedical and health-related keywords that describe the subject of an article, as well as free text terms, to expand our search. We used the tiab (title and abstract) code after each free text term to restrict our query and find the relevant articles. The exclusion criteria included articles published in languages other than English.

Search terms included the following: COVID-19, post-COVID-19 pulmonary fibrosis, idiopathic pulmonary fibrosis, histopathologic, clinical, and imaging aspects.

The search resulted in 2567 total articles. The most frequently used databases and the selection criteria represented the search strategy to retrieve the studies. An example is (search* [tiab] OR medline [tiab] OR clarivate [tiab] OR pubmed [tiab] OR [tiab] [tiab] OR google scholar [tiab] OR science direct [tiab]) AND (selection criteria [tiab] OR study selection [tiab] OR eligibility criteria [tiab] OR inclusion criteria [tiab] OR exclusion criteria [tiab]).

The authors independently reviewed the titles and abstracts included in this review. The investigation was supplemented by reviewing reference lists of included studies and related review papers.

3. Risk Factors

Recognizing the potential risk factors for PCPF is a primary goal for improving the clinical course of these patients. It has been demonstrated that PCPF has several shared significant risk factors with IPF (Table 1) [14]. Some studies from different countries showed that patients at higher risk for PCPF are elders, males, smokers, and patients with underlying conditions such as diabetes and cardiovascular and lung diseases [15–18]. Smoker patients with COVID-19 are 2.4 times more likely to need ICU admission and mechanical ventilation or die than nonsmokers [19,20]. Studies demonstrated that cigarette smoking determines endoplasmic reticulum stress; production of transforming growth factor beta (TGFB) [21], which mediates fibrosis; increased epithelial permeability; production of reactive oxidative species; and alteration of tissue regeneration, inducing lung micro-injuries [22].

Table 1. Frequent risk factors for PCPF and IPF.

Risk Factor	PCPF	IPF
Gender-male	yes	yes
Advanced age	yes	yes
Smoking	yes	yes
Comorbidities	yes	yes
Toxic environmental exposure	no	yes
Severity of dyspnea	yes	yes
Prolonged hospital stay	yes	no
ICU admission	yes	yes
Mechanical ventilation	yes	yes
Lung microbiome	yes	yes

Furthermore, different features related to the severity of the acute phase of the disease were associated with a higher risk of developing PCPF, such as the presence of dyspnea, prolonged hospital stay, intensive care unit (ICU) admission, the use of high-flow oxygen support, the need for intubation and mechanical ventilation, and development of ARDS [15,23,24]. McGroder's study has shown that at four months of follow-up, in a population with severe COVID-19, 72% of the patients who had mechanical ventilation developed PCPF, as opposed to 20% of patients who did not [25]. Opposite to the severe forms of COVID-19, patients with mild-to-moderate disease had no fibrotic abnormalities on the CT scan follow-up [26].

As the global COVID-19 pandemic has progressed, various studies have identified different markers and biomarkers that may predict the development of PCPF [17,23]. In retrospective studies from China, higher levels of serum lactic dehydrogenase and inflammatory biomarkers, including C-reactive protein (CRP), and interleukin-6 (IL-6) were found in the subgroups of patients who presented changes compatible with pulmonary fibrosis on follow-up CT scans [17,18]. Decreased lymphocyte count [27] and lower plasma levels of interferon-γ (IFN-γ) are other laboratory markers associated with a high risk of developing PCPF [28].

Regarding the risk factors for the development of IPF, there is increasing evidence to support the role played by the intrinsic risk factors (e.g., advanced age, male sex, genetics, lung microbiome), comorbidities (e.g., diabetes mellitus, gastroesophageal reflux, obstructive sleep apnea, herpes virus infection), and extrinsic risk factors (e.g., smoking, air pollution, environmental exposures). These risk factors, some of them in common with those predisposing to PCPF, may independently increase susceptibility for IPF or work in a synergistic mode to contribute to a higher risk for disease development [29] or be associated with shorter survival rates [30]. The incidence and prevalence of IPF increase with age, being higher in adults 65 years or older. The mechanisms include altering the proliferation/apoptosis ratio [31], decreasing alveolar stability, and reducing differentiation capacity, which promotes fibrosis [32].

Around 70% of all IPF patients are male [33], males also being 1.3 times more subjected to PCPF than females [34]. Studies performed on animals gave a possible explanation, showing that female hormones have a role in protection against pulmonary fibrosis [35,36]. Another cause may be that men are more exposed to tobacco smoking and occupational exposures [33]. Smoking history increases the risk of developing IPF by 60% [37] and the risk of progression to severe forms of COVID-19 by 1.3 times [4]. The environmental and occupational factors associated with IPF are metal dust, wood dust, stone, sand, and farming substances [38].

Genetic predisposition is also a significant risk factor for PCPF and IPF (Table 2). Host genetic predisposition was proposed as a risk factor for severe courses of COVID-19 [39]. The first genome-wide association studies (GWASs) identified the 3p21.31 gene cluster (rs11385942) associated with severe forms of COVID-19 and respiratory failure and confirmed a potential involvement of the ABO blood group system [40]. The COVID-19 Host Genetics Initiative (HGI) meta-analysis also confirmed the 3p21.31 locus as significant, with two signals within the locus, one associated with severity (rs10490770) and the other with infection susceptibility (lead variant rs2271616) [41]. CXCR6 recruits CD8-resident memory T cells in the respiratory tract to combat respiratory pathogens and is a causal gene for severe disease (lead variant rs10490770) [42].

Different studies reported that several genes are associated with IPF predisposition, including the genes that encode the surfactant and proteins A and C (SFTPC, SFTPA1, SFTPA2), genes associated with telomerase dysfunction (TERT, TERC, DKC1, RTELI, PARN), genes affecting the integrity of the epithelial barrier (DSP), and genes affecting host defense (MUC5B, TOLLIP) [43,44]. The most recent meta-analysis reported an association between MUC5Brs35705950 and hospitalization due to COVID-19 [45].

The study of Fadista showed that genetic variants associated with IPF did not predispose to an increased risk of severe COVID-19 [14]. As mucins are involved in the first defense against pathogens in the airways and play an essential role in mucociliary clearance, high expression may protect against SARS-CoV-2 infection. Nevertheless, this study was driven by a single abnormal variant at the MUC5B locus, which had an apparent protective effect on the severity of COVID-19. Removal of this outlier demonstrated that the remaining variants associated with increased risk of IPF were also associated with increased risk of severe COVID-19.

Table 2. Genetic mutations in IPF and PCPF.

Gene	Mutation Consequence
SPTPC	Altered encoding of surfactant proteins
SPTPA1	Altered encoding of surfactant proteins
SFTPA2	Altered encoding of surfactant proteins
TERT	Telomere shortening
TERC	Telomere shortening
DKC1	Telomere shortening
RTELI	Telomere shortening
PARN	Telomere shortening
DSP	The affection of the epithelial barrier
MUC5B	Host defense affection
TOLLIP	Host defense affection

Studies have shown that microbial pathogens such as Streptococcus and Staphylococcus are associated with the development and progression of IPF [46,47]. At the same time, COVID-19 patients with bacterial infections were more frequently admitted to the ICU and needed invasive ventilation [48]. A recent meta-analysis [49] showed a prevalence of bacterial co-infections of 3.5% in COVID-19 patients admitted to the ICU, with most of the cases being hospital-acquired infections with Gram-negative germs, complicating the evolution of the disease, prolonging the hospitalization time, and increasing the risk of mechanical ventilation, all of which are risk factors for PCPF development.

In conclusion, PCPF and IPF have common risk factors such as advanced age, male sex, smoker status, comorbidities, and genetic mutations. At the same time, toxic environmental exposure is a risk factor incriminated in IPF, not proven yet in PCPF.

4. Clinical Aspects

The clinical manifestations of PCPF and IPF are similar (Table 3), with both diseases sharing numerous symptoms, including dyspnea, dry cough, fatigue, chest pain, and weight loss, which are related to decreased life quality [50,51]. There is a lack of data for the clinical course of PCPF, but the results of the relevant prospective studies will fill this gap.

Table 3. Clinical aspects in PCPF and IPF.

Clinical Aspect	PCPF	IPF
Dyspnea	yes	yes
Cough	yes	yes
Fever	yes	no
Fatigue	yes	yes
Chest pain	yes	yes
Depression	yes	yes
Velcro crackles	yes	yes
Wheezing	yes	no
Clubbing fingers	no	yes

A study conducted by Farghaly assessed persistent symptoms in patients with PCPF at six-month follow-up after the acute disease and showed that the most common symptoms are dyspnea (98%), dry cough (91%), fever (70%), productive cough (19%), and chest pain

(16%). The complications of PCPF that increase the risk of death are respiratory failure, sepsis, and acute kidney injury [52].

Another prospective cohort study of 76 severe COVID-19 patients requiring supplemental oxygen found a positive correlation between the presence of radiographic abnormalities (fibrotic-like patterns) 4 months after hospitalization and decreased lung function, cough, and frailty [25].

The study conducted by Kamal showed that almost 90% of the 287 included patients suffered from several symptoms and diseases after recovery from COVID-19. Most individuals suffered from fatigue (72.8%), anxiety (38%), joint pain (31.4%), continuous headache (28.9%), chest pain (28.9%), dementia (28.6%), depression (28.6%) and dyspnea (28.2%), while 2.4% of recovered patients were newly diagnosed with diabetes [53]. There was a positive correlation between the initial disease's severity and post-COVID-19 manifestations, many related to the central nervous system, including continuous headache, anxiety, depression, and obsessive–compulsive disorder.

Psychological morbidities such as anxiety and depression were also reported in IPF patients [54]. Anxiety and depression are strongly associated with health-related quality of life [55]. For IPF patients and subjects recovering from COVID-19, continuous counseling is essential to detect warning signs of developing severe manifestations and maintain good medication adherence.

A case–control study performed on the UK population in 2017 has shown a strong correlation between dyspnea and cough as debut symptoms in IPF, appearing up to four years before the diagnosis [56].

Obstructive sleep apnea is also highly associated with IPF, with patients reporting snoring, insomnia, daytime sleepiness, and witnessed apneas. Gastro-esophageal reflux can be present in 35% to 100% of patients with IPF. These patients can be asymptomatic or present different digestive symptoms, such as regurgitation, belching, dysphagia, dysphonia, and chest pain. Cough can be associated with about 28% of cases of gastro-esophageal reflux [57].

The literature data have well documented the physical findings of IPF patients compared to PCPF patients. A study analyzing patients with PCPF six months after an acute episode has shown the presence of pathological auscultation sounds in 4% to 12% of the patients, represented by Velcro crackles and wheezing [58].

Fine crackles, usually in the lower posterior parts of the lung, are typically reported in IPF patients, while clubbing fingers are found in 30–50% of the cases, correlated with smooth muscle proliferation in the areas of fibrosis observed in lung biopsy. Body mass index also correlates with IPF patients' survival [59].

In conclusion, PCPF and IPF have similar manifestations, such as dyspnea and cough, and physical findings represented by Velcro crackles, both evolving in time with respiratory failure, increasing the death risk in these patients.

5. Pulmonary Function Tests

Pulmonary function tests (PFTs) indicate a restrictive pattern and an altered lung diffusion capacity for carbon monoxide (DLCO) in the vast majority of cases of patients who have recovered after severe forms of COVID-19 (Table 4) [60,61]. In a recently published meta-analysis, which included 380 post-COVID-19 patients, the authors found impaired DLCO and restrictive and obstructive patterns in 39%, 15%, and 7% of subjects [61]. A high prevalence of decreased DLCO (66%) was found in patients with severe COVID-19, especially those with elevated inflammatory markers, who were more likely to develop pulmonary fibrosis [18]. The British Thoracic Society (BTS) guide suggests evaluating patients with severe COVID-19 with full PFTs 12 weeks after hospital discharge [62]. According to the findings of Cherrez-Ojeda, the impairment in PFTs appears to persist well beyond this timeframe [63]. A prospective observational study that analyzed the evolution of functional and radiological features between 3 and 6 months after hospital discharge in critical COVID-19 survivors reported the persistence of functional abnormalities such as impairments in total lung capacity (TLC) (41% and 33%) and DLCO (88% and 80%) at the end of the monitoring period [64]. Significant improvements were observed only in forced expiratory volume in 1 second (FEV1), forced vital capacity (FVC), and distance covered during the 6-minute walking test (6MWT).

Table 4. PFTs in PCPF and IPF.

PFT	PCPF	IPF
Restrictive dysfunction	yes	yes
Obstructive dysfunction	yes	Coexistence of obstructive disease
Decreased DLCO	yes	yes
6MWT desaturation	yes	yes

Acute phase severity markers such as the presence of previous conditions (arterial hypertension and diabetes), invasive mechanical ventilation (IMV), and prone positioning during the ICU stay were associated with a low level of DLCO and its non-improvement during the 3rd and 6th months of follow-up. The decrease in DLCO could result from interstitial or pulmonary vascular abnormalities caused by critical COVID-19. The authors observed that despite the partial radiological resolution, gas–blood exchange abnormalities persisted at the six-month follow-up, which could suggest the initial establishment of an irreversible, chronic lung disorder.

Although the restrictive pattern is consistent also in IPF, with reduced FVC and TLC in one study, 25% of the patients had normal values of TLC, and more than 50% had normal FVC (Table 4). The obstructive pattern is not typical in IPF, suggesting the coexistence of an obstructive cause. The lung function decline is a mortality predictor in IPF; as the restriction's severity increases, FEV_1/FVC increases, and DLco decreases. Patients with a drop bigger than 10% in FVC in 6 months or desaturate less than 88% at the 6MWT have a higher mortality risk [65]. In IPF, the decrease in DLCO usually precedes restrictive dysfunction, and DLco lower than 35% or a decline of more than 15% in one year is also associated with high mortality [66,67].

In conclusion, because the changes in forced vital capacity (FVC), total lung capacity (TLC), and DLCO are predictive factors for mortality in IPF patients, long-term monitoring through PFTs may also be justified in patients with PCPF.

6. Radiologic Aspects

In the early stages of COVID-19, the most common radiological findings are bilateral "ground glass opacities" (GGOs) and consolidations, predominantly in the lower lobes, posterior and peripheral. The radiologic exam may also show a "crazy-paving" pattern, nodular opacities, halo sign, reversed halo sign, pleural effusions, cavitation, and lymph node enlargement [15,68,69]. The extent of the lesions varies significantly; they can be patchy or diffuse, and a predominance of central and upper distribution may also be present. Different pulmonary CT abnormalities may be found even in patients without respiratory symptoms.

The lesions from the acute phase of the disease may progress to fibrotic abnormalities such as interlobular septal thickening and traction bronchiectasis, especially in survivors of critical forms of COVID-19 [70]. However, the data regarding the long-term evolution of pulmonary changes in these patients are scarce. Liu D reported that between three and four weeks after the acute COVID-19 pneumonia, a transitory extension of the GGOs occurs, with a decrease in density, an aspect described as "tinted sign", accompanied by the distortion of the bronchovascular bundle [7]. Most patients with mild or moderate pneumonia have complete resolution of the imaged lesions in the first month after the episode; the first lesions that resolve are GGOs. Other lesions such as subpleural bands and bronchial dilations are remitted slowly, the severity of the initial clinical manifestations being the most common determinant of the resolution time.

Regarding the severity of lung parenchymal affection, the most commonly used CT score has a range from 0 to 25, each lobe affection being visually scored on a scale of 0–5, with 0 indicating no involvement; 1, less than 5% involvement; 2, 5–25% involvement; 3, 26–49% involvement; 4, 50–75% involvement; and 5, more than 75% involvement [71].

Recent studies evaluated long-term longitudinal changes in chest CT findings in COVID-19 survivors. Yin X showed that GGOs remained on CT images in 30% of the subjects more than six months after discharge [72]. At the same time, reticulation was observed in 46% of cases, more commonly after severe forms of the disease, with a longer duration of hospitalization. In a one-year follow-up study, the CT scans were normal in 16% of cases, were stable in 19%, and showed a reduction in lesion extension in 65% of the patients, in reticular band opacities persisting at one-year follow-up. At the same time, the extent of GGOs was substantially reduced, and the long-term persistence of CT abnormalities after COVID-19 was correlated with the disease's severity, high inflammatory state, and the need for more intensive ventilator requirements [73].

According to the CT scoring method proposed by Camiciottoli, the degree of pulmonary fibrosis was evaluated by ground glass opacity, linear opacity, interlobular septal thickening, reticulation, honeycombing, or bronchiectasis features in chest CT images [74]. In the study of Jia-Ni Zou, approximately 80% of the 284 COVID-19 patients had pulmonary fibrosis at discharge, and it was more pronounced in patients with severe disease than in those with mild/moderate form (73.8%).

Farghaly showed the presence of GGOs (95%), honeycombing (25%), and pulmonary consolidations (9%) in patients with PCPF. The CT score was higher for patients with mechanical ventilation or ICU admission. A high CT score was also associated with prolonged hospitalization and severe dyspnea [52].

PCPF changes appear in the areas where there were previously GGOs during COVID-19 pneumonia [6]. As a result, the CT distribution of the fibrotic changes will be predominantly bilateral, peripheral, and in the lower lobes. A similar distribution of CT abnormalities was also described in IPF cases, in which usual interstitial pneumonia (UIP) represents the hallmark radiologic CT pattern (Figure 1).

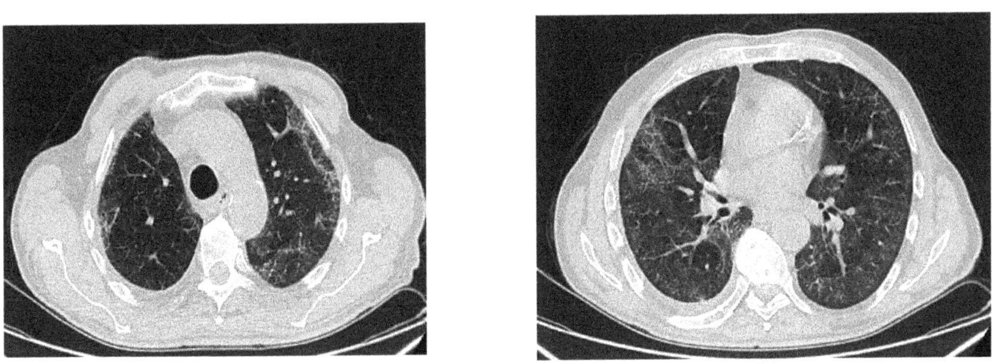

(a) (b)

Figure 1. *Courtesy of Dr Oana Cristina Arghir, who provided images of post-COVID-19 pulmonary fibrosis from Pneumology Hospital of Constanta, Romania.* Thoracic CT scan of a patient with PCF, one year after the acute episode, showing subpleural and peribronchovascular reticular opacities, traction bronchiectasis (**a**), and GGO bilateral and fibrotic lines predominantly in the upper lobes (**b**), in a 77-year-old male, with history of severe COVID-19 in October 2021 and PCPF in October 2022.

Regarding the high-resolution computer tomography (HRCT) findings, the presence of honeycombing is the main difference between IPF and PCPF, being essential for the definite diagnosis of IPF but rare in PCPF, with just a few cases being reported [75]. Another chest CT feature that differentiates those two diseases is the large extent of the GGOs in

PCPF, while in IPF, GCOs are commonly absent or minimal or may be present in case of exacerbations. The evolution of pulmonary manifestation on CT scan is very well documented in IPF, and the data from the literature show an irreversible, progressive course [76,77], while despite the scarce data regarding the evolution of post-COVID-19 CT abnormalities, recent studies found an improvement, with a 10–40% decrease in the extent of CT lesions and no progression at 1-year follow-up CT [72,74]. Prospective studies from large cohorts undergoing more prolonged monitoring are likely further to clarify the evolution of PCPF on CT scans.

In 2011, the clinical practice guidelines for the diagnosis of IPF from the American Thoracic Society/European Respiratory Society/Japanese Respiratory Society/Latin American Thoracic Association (ATS/ERS/JRS/ALAT) proposed the concept of multidisciplinary diagnosis for IPF in patients without surgical lung biopsy (SLB) if the patient had a UIP pattern on HRCT and suggestive clinical presentation, and recommend that HRCT features of IPF should be referred to as UIP, possible UIP, or inconsistent with UIP. The imaged UIP pattern is an accurate indicator of the presence of histopathological UIP patterns.

The UIP pattern consists of reticular opacities, honeycombing, traction bronchiectasis, and bronchiolectasis, observed with fine reticulation, and, in the case of exacerbations, GGOs. Honeycombing is a defining feature of UIP, mandatory for a definite diagnosis, associated or not with peripheral traction bronchiectasis or bronchiolectasis (Figure 2a,b).

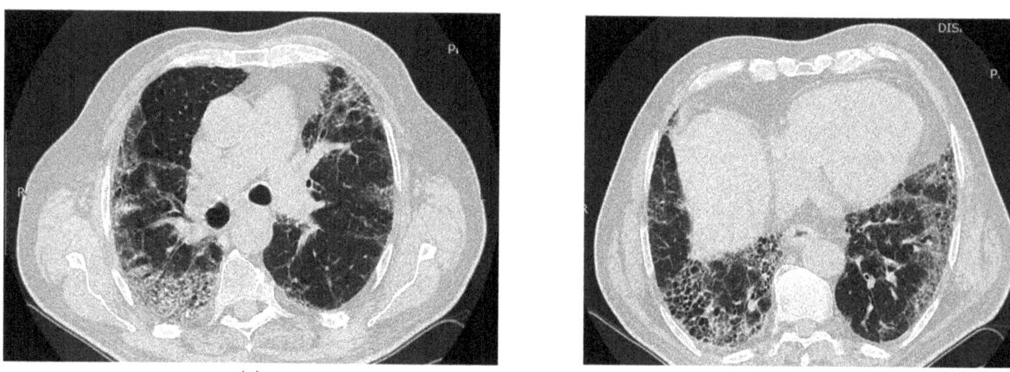

(a) (b)

Figure 2. *Courtesy of Dr Ariadna Petronela Fildan, who provided images of IPF from Pneumology Hospital of Constanta, Romania.* Thoracic CT scan of a patient with IPF showing honeycombing and reticular opacities with the basal and subpleural distribution in a 73-year-old male smoker, with exposure to environmental pollutants (**a**), with idiopathic pulmonary fibrosis (**b**), who died in 2017, 3 years after the diagnosis.

Except for honeycombing, UIP features on the HRCT define the "possible UIP" pattern; in this case, a lung biopsy is necessary for the positive diagnosis. The presence of other abnormalities, such as extensive GGOs, micronodules, pleural abnormalities, non-honeycombing cysts, consolidation areas, or the predominance of peribronchovascular or perilymphatic distribution, is characteristic of inconsistency with the UIP pattern. However, differentiating honeycombing from traction bronchiectasis and emphysema can be difficult. Honeycombing has a thicker wall and subpleural distribution parallel with the chest's border, while emphysema has cystic airspaces with thin walls located further away from the chest wall [78].

In conclusion, fibrotic modifications have a basal and subpleural preponderance in PCPF and IPF. Honeycombing, mandatory for the UIP pattern and diagnosis of IPF, is only found in 25% of PCPF cases, while GGOs with extensive distribution in PCPF are present only in IPF exacerbation.

7. Histopathologic Characterization

The histopathological data are mainly based on autopsy findings. In contrast to other viral infections such as H1N1, for which death occurred within a few days after the symptomatic debut, the patients infected with SARS-CoV-2 died within a mean duration of three weeks. An explanation may be the more progressive lung injury caused by SARS-CoV-2 that favors the repair with extracellular marker deposition. However, histopathological studies in prolonged or post-COVID-19 patients are limited because PCPF is a relatively new entity, and few studies using autopsies were performed. This is an essential factor that needs to be further considered because this lack of knowledge also limits the etiopathogenic understanding of the disease and the development of specific therapy.

The autopsy studies show that during the course of the disease, type I and type II pneumocytes are infected by the SARS-CoV-2 virus, resulting in a cytopathic effect, pneumocyte desquamation, accumulation of fibrinoid material in alveolar spaces, and numerous inflammatory cells in the lungs, macrophages, lymphocytes, and neutrophils [79].

Infected type II pneumocytes contain numerous autophagosomes, ultrastructurally characterized by double membranes and the presence of organelles in the cytoplasm, also containing viral aggregates, that can be present in tracheal epithelial cells and within the extracellular mucus in the tracheal lumen. Immunohistochemical staining can demonstrate virus particles' existence using monoclonal antibodies against the SARS-CoV-2 nucleocapsid protein [80]. All these changes represent the spectrum of diffuse alveolar damage (DAD), which is characteristic of the acute or exudative phase of severe COVID-19, with numerous reactive pneumocytes, lung hemorrhage, fibrin deposits in the alveolar spaces, interstitial edema, hyaline membranes, giant cell formation, and bronchiolitis obliterans [81–84]. Thrombotic events in pulmonary arteries may also occur in this phase.

DAD results from acute lung injury (ALI) determined by direct or indirect causes, or in the case of ARDS, requiring mechanical ventilation. ARDS is defined as acute hypoxemia with a ratio of partial pressure of arterial oxygen to the fraction of inspired oxygen (PaO_2:FiO_2) of a maximum of 200 mmHg. At the same time, ALI, being less severe, refers to acute hypoxemia with a PaO_2:FiO_2 ratio of 300 mg Hg [85]. Other histopathologic findings in the case of ARDS and ALI are acute eosinophilic pneumonia (AEP) and acute fibrinous and organizing pneumonia (AFOP) [86].

DAD has two phases: the acute phase in the first week after lung injury and the organizing or proliferative phase [87]. Two days after the lung injury, hyaline membranes are developed, while thrombi can also be seen as a result of the alteration of the coagulation without any underlying thromboembolic disorder [88]. The organizing phase is defined by cellular fibroblastic proliferation, type 2 pneumocyte hyperplasia, squamous metaplasia, and residual fibrin rest. In this phase, the hyaline membranes become integrated into the alveolar septa and cannot be seen anymore [89]. In autopsies, other modifications such as polypoid plugs, alterations of the basal membrane, and thickening of the alveolar wall were also present [82,89,90]. After this phase, DAD can be resolved gradually or evolve into interstitial fibrosis (fibrosing stage) (Figure 3a,b), with the excessive extracellular matrix, dense collagen deposition, and diffuse thickening of alveolar walls, resulting in an architectural disorder similar to other cellular and fibrous interstitial pneumonia patterns [89,91,92]. The most common finding in 30 minimal invasive autopsies was organizing DAD (70%), acute DAD (40%), and/or fibrosing patterns. Fibrosing DAD may be involved in the development of post-COVID-19 pulmonary fibrosis. There are still limited data about the pathology of prolonged disease [89]. The study performed by Hanley B on ten autopsies reported DAD presence in all the cases, as well as lymphocyte inflammation, predominantly CD4+-positive T cells, along with macrophages and scattered plasma cells. Chronic bronchiolitis was a common finding, and CD56-positive natural killer cells were rare (Figure 3b). Thromboembolism was a frequent finding in small and medium-sized vessels, without any sign of deep venous thrombosis in the external examination [93].

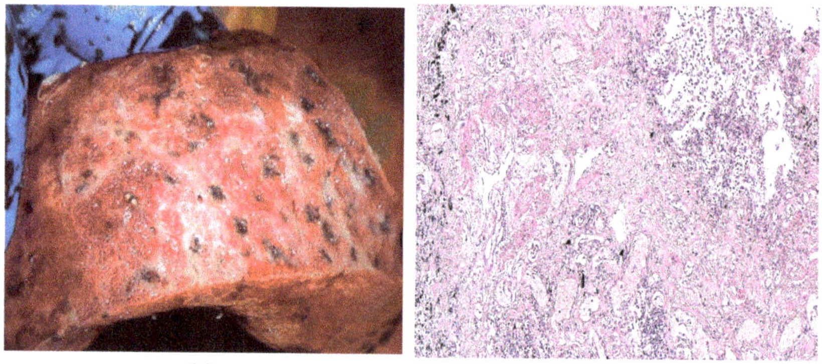

(**a**) (**b**)

Figure 3. *Courtesy of Dr Mariana Deacu, who provided images of post-COVID-19 pulmonary fibrosis from "St. Andrew" Emergency County Hospital of Constanta, Romania.* (**a**) Macroscopic and (**b**) microscopic aspect of the lung in a 55-year-old male patient with COVID-19 interstitial pneumonia in the fibrosing stage, whose autopsy was performed three months after SARS-CoV-2 infection, showing architectural disorder caused by excessive extracellular matrix, dense collagen deposition, and diffuse thickening of alveolar walls. HE ×40.

The histological patterns observed in organizing pneumonia are represented by fibroblasts and myofibroblasts that fill the alveolar space and ducts as an inflammatory response to the virus. This mechanism may have a favorable evolution towards resorption or may chronically evolve to excessive collagen deposition and, thus, pulmonary fibrosis [94]. In explanted lungs from patients with lung transplants due to COVID-19, the main pathologic characteristic was extensive pulmonary fibrosis, acute interstitial pneumonia or organizing pneumonia, micro-thrombosis, alveolar hemorrhage, and acute bronchopneumonia from superimposed bacterial infection [95].

The morphologic changes of the lung in COVID-19 are the following: (1) reactive epithelial changes and DAD; (2) vascular with microvascular damage, microthrombi, and acute fibrinous and organizing pneumonia; (3) fibrotic changes, with evidence of interstitial fibrosis [96]. The epithelial and vascular changes appear alone, simultaneously, or consecutively in all stages. Buja described the association of DAD with microvascular involvement and proposed three stages of lung injury: early infection stage, pulmonary stage, and severe hyper-inflammation stage. In the first stage, the morphologic aspects are of interstitial pneumonia with DAD, in some cases associated with micro-thrombosis, peripheral lung hemorrhage, and, in severe cases, pulmonary thromboembolism [97]. In COVID-19, interstitial inflammatory infiltrate is reduced (Figure 4a,b), unlike typical interstitial pneumonia.

Another study by Ackerman highlights the presence of thrombosis (Figure 5a,b), pulmonary vascular endothelitis, and angiogenetic alterations in patients with COVID-19 [98]. In contrast, Burel's study showed the loss of pericytes, the cells responsible for micro-vessel integrity, which may trigger micro-vasculopathy [99]. Ackerman's study has described three angiogenetic features of COVID-19, the first being severe endothelial injury associated with the destruction of the endothelial cell membranes, the second being disseminated vascular thrombosis in the lungs, associated with microangiopathy and capillaries occlusion, and the third feature describing new vessel growth in the lungs through angiogenesis [98]. One study has shown an increased number of angiotensin-converting enzyme 2 (ACE2)-positive cells in the lungs of COVID-19 patients. SARS-CoV-2 within the endothelial cells suggests perivascular inflammation, the direct effects of the virus causing endothelial injury [100].

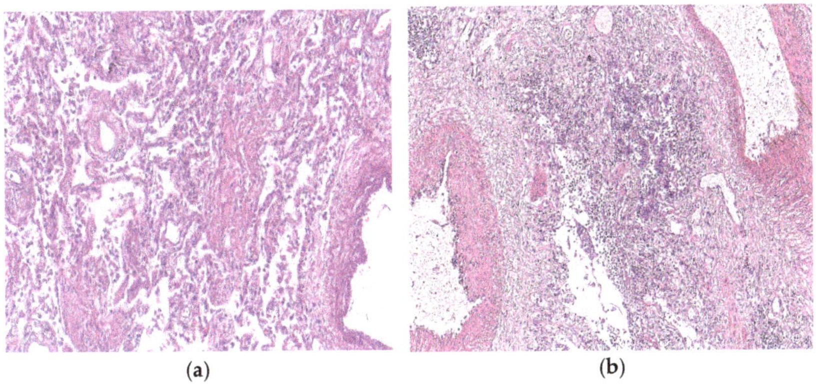

Figure 4. *Courtesy of Dr Mariana Deacu, who provided images of post-COVID-19 pulmonary fibrosis from "St. Andrew" Emergency County Hospital of Constanta, Romania.* Microscopic view of the lung in a 68-year-old patient with COVID-19, whose necropsy was performed five weeks after SARS-CoV-2 infection, showing (a) interstitial inflammatory infiltrate reduced and (b) chronic bronchiolitis. HE ×100.

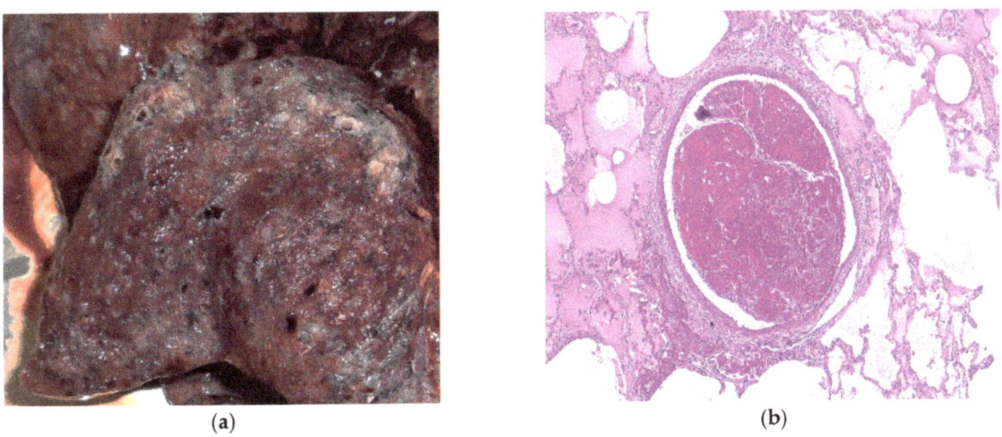

Figure 5. *Courtesy of Dr Mariana Deacu, who provided images of post-COVID-19 pulmonary fibrosis from "St. Andrew" Emergency County Hospital of Constanta, Romania.* (a) Macroscopic and (b) microscopic aspects of the lung in a 62-year-old patient showing thrombosis, whose necropsy was performed three weeks after SARS-CoV-2 infection, showing pulmonary vascular endothelitis and angiogenetic alterations. HE ×40.

Some authors suggest two patterns of fatal COVID-19, with different clinical courses: one with high viral load and high cytokine expression in the lung but a limited morphologic expression, and a second one with low viral load and cytokine expression but a large number of immune cells (including CD8 + T lymphocytes and macrophages), which correlates with the presence of DAD [101].

DAD's proliferative/organizing phase shows type II pneumocyte hyperplasia, reactive pneumocytes (Figure 6a), alveolar wall thickening, and myofibroblast proliferation (Figure 6b). A case report of an 80-year-old woman with subsequent negative SARS-CoV-2 at the time of the autopsy showed severe reactive and inflammatory changes in all the lung samples. The architecture was destroyed in larger areas with fibrinous organization and collagenized fibrosis. Widespread angiogenesis was seen, along with focal bleeding. Local moderate chronic inflammation dominated by lymphocytes was present. The fibrosis had a honeycomb-like pattern with enlarged airspaces and bronchial metaplasia in some areas. A small subpleural area showed alveoli with hyaline membranes representing the acute stage of lung injury, as is seen in acute DAD.

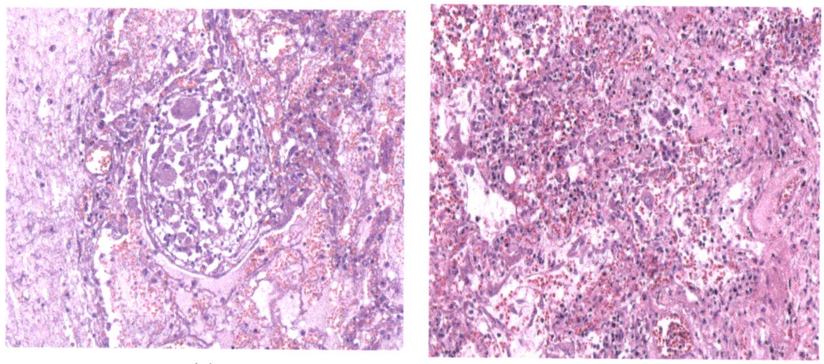

(a) (b)

Figure 6. *Courtesy of Dr Mariana Deacu, who provided images of post-COVID-19 pulmonary fibrosis from "St. Andrew" Emergency County Hospital of Constanta, Romania.* Microscopic view of the lung in a 68-year-old patient, whose necropsy was performed eight months after SARS-CoV-2 infection, showing (**a**) type II pneumocyte hyperplasia and reactive pneumocytes, (**b**) alveolar wall thickening, and myofibroblast proliferation. HE ×100.

The guideline panel updated the diagnostic criteria for idiopathic lung fibrosis and defined patterns of UIP, probable UIP, indeterminate UIP, and alternate diagnosis. To confirm those cases, they recommended performing bronchoalveolar lavage (BAL) and surgical lung biopsy.

The 2018 ATS/ERS/JRS/ALAT guideline classifies the histopathological findings into UIP, probable UIP, and indeterminate UIP. IPF is macroscopically characterized by honeycombing, consisting of fibrosis in the interior part of the lobes, predominantly subpleural. The microscopic aspect of the UIP pattern is composed of patchy dense fibrosis, with architectural distortion, predominantly in the periphery of the lobule and paraseptal, with a regular central portion of the lobule. As a result, the patterns will evolve from chronic to acute to absent from the periphery to the centum of the lobule. The honeycombing aspect can be seen subpleural as dense fibrosis surrounding bronchial epithelium, which lines irregular airspaces in the centrum of the lobule, where significant inflammation and fibrosis are absent. Fibroblast foci can be seen in the area between regular regions of the lobule and fibrotic lesions, arranged parallel with the alveolar foci, on a basophilic myxoid ground (Figure 7a,b).

Some histological UIP features characterize a probable UIP pattern to the extent that precludes the definitive diagnosis, the absence of elements of an alternative diagnosis, or the presence of honeycombing alone. Indeterminate UIP is characterized by fibrosis with or without architectural distortion, suggesting UIP secondary to other causes or non-suggestive for UIP or features of UIP, along with elements of an alternative diagnosis. An alternative diagnosis is indicated by histological aspects suggestive of other diseases at biopsy, such as airway-centered lesions, granulomas, interstitial inflammation without fibrosis, chronic fibrous pleuritis, and hyaline membranes [102].

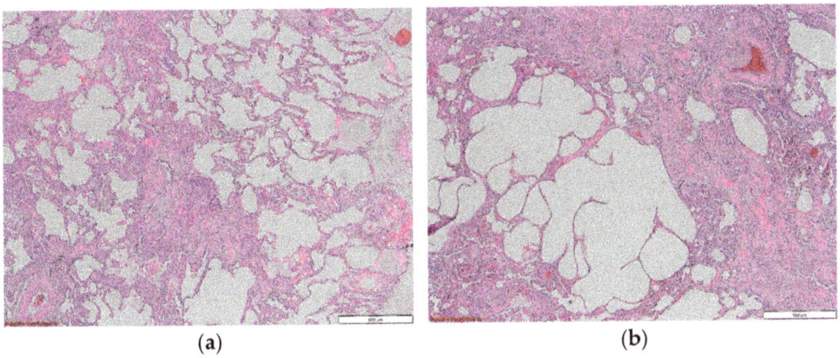

Figure 7. *Courtesy of Dr Angela-Ştefania Marghescu, who provided images of IPF from Pneumology Institute "Marius Nasta" Bucharest, Romania.* (**a**) Spatial variability, with normal lung parenchyma mixed with fibrotic areas (fibroblastic focus); HE ×40. (**b**) Interstitial fibrosis with architectural distortion and cystic changes in the pulmonary parenchyma; HE ×40, in a 62-year-old woman, whose necropsy was performed three months after SARS-CoV-2 infection, with marked pulmonary fibrosis.

The histopathological guidelines for UIP were changed in 2021 and require advanced fibrosis with distortion of the architecture, beginning at the periphery of the lobule and going to the centrilobular regions. At this level, fibroblast foci are often encountered as evidence of active injury, usually situated at the interface between fibrotic and non-fibrotic areas, without features characteristic for an alternative diagnosis. A lack of inflammation and ununiform affection of lung parenchyma characterize UIP.

In acute exacerbations of UIP, an ALI pattern can be seen, predominantly in the regions without chronic fibrosis, while the peripheric fibrotic lesions are unmodified. In the centrum of the lobule, we can see modifications of DAD, such as type II pneumocyte hyperplasia, diffuse alveolar septal thickening caused by edema, accumulation of airspace fibrin or nonspecific changes such as thrombi within small pulmonary arteries, distal airway squamous metaplasia, and the presence of hyaline membranes [78].

In acute exacerbation of IPF, we can find DAD and UIP features, so DAD is a histopathological feature in IPF and COVID-19. Numerous factors, such as infections, shock, sepsis, connective tissue disorders, and disseminated intravascular coagulation, can cause DAD. When the etiology is unidentified, it is called acute interstitial pneumonia, previously referred to as Hamman–Rich syndrome [103].

In conclusion, IPF and PCPF both cause DAD and lung fibrosis. The data regarding PCPF are currently limited, and further studies with histopathological examination are needed.

8. Therapeutic Perspectives

In COVID-19, cytokine storms, oxidative stress, and inflammation are involved, so the proposed therapy for PCPF consists of anti-fibrotic and anti-inflammatory drugs.

Pirfenidone (a pyridine) and nintedanib (a tyrosine kinase inhibitor) are anti-fibrotic drugs used in IPF that reduce lung function decline by 50% and improve life expectancy by 2–5 years [104]. Neither of these drugs has an immunosuppressive effect, so they should not be stopped in case of infections.

Since April 2020, these drugs have been available exclusively in oral form. As a result, it is impossible to administrate them in the case of mechanically ventilated and intubated patients, such as patients with severe COVID-19. A form of pirfenidone with inhalator administration is under evaluation for COVID-19 patients. Pirfenidone should not be administrated in patients with a glomerular filtration rate of less than 30 mL/min per 1·73 m^2. Patients with severe COVID-19 are at high risk of developing renal dysfunction, so pirfenidone should be carefully considered in these patients.

Pirfenidone and nintedanib can determine hepatotoxicity, while liver function test alterations are commonly found in severe COVID-19, so in this case, temporary disruption of the anti-fibrotic treatment might be necessary [105].

Nintedanib has an increased risk of bleeding in case of concomitant administration with a full dose of an anticoagulant. COVID-19 patients have an increased risk of acute pulmonary embolism, and anticoagulant therapy is necessary. Anti-fibrotic treatment for COVID-19 might be helpful to patients with a poor prognosis and a high risk of developing pulmonary fibrosis and ALI [106].

The SARS-CoV-2 spike protein has an Arg-Gly-Asp integrin-binding domain raising the possibility that inhibitory integrin or galectin therapies might be helpful as COVID-19 treatment. Some drugs in development can target molecules from TGF-β pathways, such as those against αvβ6 integrin (BG0001, PLN-74809) and galectins (TD139). Studies on mice have shown that those who did not have αvβ6 integrin or had treatment with an αvβ6 blocking antibody had increased protection against viral infections [107]. In another study on mice, galectin Gal-3 determined reduced pulmonary inflammation and protection against TGF-β-induced lung injury and fibrosis [108].

mTOR is a target in IPF, and two recent studies showed that mTOR might be an anti-SARS-CoV-2 target, with rapamycin being considered in COVID-19 patients [109].

Pentraxins are response proteins of the acute phase, with a role in immunity and inflammation. PRM-151, an analog of SAP (PTX2), has shown promising results in IPF trials. SAP determines suppression of JNK family signaling, as a JNK1 inhibitor preventing fibrosis 55 and inhibiting ALI [110].

C21 role (an agonist of AT2R) is studied for COVID-19 and has clinical trial applicability in IPF, having anti-inflammatory properties. ACE2 receptors are the primary SAR-CoV-2 receptors. A study had shown that in patients who took AT1R blockers before hospitalization, the risk of severe COVID-19 was significantly decreased [111].

Treamid or bisamide derivative of dicarboxylic acid (BDDA) is an experimental drug with promising results used in animals with pulmonary fibrosis that inhibits the production and deposition of collagen, being in trial for use in cases of IPF and post-COVID-19 fibrosis [112].

LYT-100 (deupirfenidon) is an N-aryl-pyridone derivative, an analog of pirfenidone, which has an anti-fibrotic effect and is in trial for use in cases of COVID-19 and IPF [113].

Corticosteroids can be used in COVID-19 and IPF exacerbation. Long-term use of corticoid therapy might reduce the risk or severity of PCPF in rats, with IPF slowing down fibrosis progression [114].

An alternative therapy that might also prevent lung fibrosis is azithromycin, a broad-spectrum macrolide, which has also proved to have antiviral and immunomodulatory effects. For these reasons, it was suggested that this potential therapy for COVID-19 can reduce the risk of PCPF [115].

Histone deacetylase inhibitors can be an alternative therapy in PCPF, with increased activity of histone deacetylase promoting the activity of TGF-ß, leading to collagen synthesis and fibrosis.

Biochanin A (isoflavone) is believed to target TGF-ß-induced fibrosis and can be an alternative therapy considered in PCPF which needs further study. A study showed a significant decrease in TGF-ß expression and collagen deposition in the lungs of the mice treated with Biochanin A [116].

Pulmonary rehabilitation, which includes exercise training, education, and behavioral changes, can improve physical and psychological conditions in cases of pulmonary fibrosis [117].

On the other hand, oxygen therapy is beneficial in IPF in patients with severe oxygen desaturation during exercise or resting hypoxemia, improving their symptoms and quality of life. Consequently, oxygen therapy is also an essential part of PCPF treatment [118].

In conclusion, there is a lack of effective specific treatment for PCPF, with numerous drug trials being studied; two are represented by nintedanib and pirfenidone, which are used for IPF and also have promising results in the treatment of PCPF.

9. Conclusions

There are numerous similar features between PCPF and IPF regarding the clinical aspects, risk factors, pulmonary function tests, and imaging and histopathological aspects, as well as some notable differences. As for clinical aspects, both pathologies have common symptoms such as dyspnea and cough; functionally, restriction syndrome and a decrease in DLco can be observed; radiologically, we can find signs of fibrosis in both cases, the main lesions being GGOs in PCPF, while the aspect is defined as UIP in IPF. Regarding histopathology, findings of lung fibrosis in COVID-19 are limited, and PCPF causes significant, irreversible consequences that affect patients' quality of life after SARS-CoV-2 infection. Further studies with histopathological examination are needed. DAD appears in PCPF but can also occur in an IPF exacerbation alongside UIP features. More studies need to be done to determine specific effective PCPF therapy. Currently, nintedanib and pirfenidone used for IPF treatment are studied as anti-fibrotic agents for PCPF.

This review would help clinicians, pathologists, and researchers better understand the mechanisms of fibrosis, make a diagnosis as accurate as possible, and help identify the patients who can be selected for anti-fibrotic therapies and future therapeutic perspectives.

Author Contributions: Conceptualization, R.-E.C. and A.-P.F.; methodology R.-E.C., M.D., M.E., G.I.B. and A.A.N.; software, A.-Ș.M.; validation, M.D., M.E., D.-E.T. and A.-P.F.; formal analysis, R.-E.C., I.G. and O.C.A.; investigation, R.-E.C., I.G., A.-Ș.M. and M.D.; resources, R.-E.C., I.G., A.-Ș.M. and M.D; data curation, R.-E.C. and I.G.; writing—original draft preparation, R.-E.C., I.G. and A.-P.F.; writing—review and editing, M.D., G.I.B., A.A.N. and O.C.A.; visualization, D.-E.T.; supervision, A.-P.F. All authors have read and agreed to the published version of the manuscript.

Funding: This research received no external funding.

Institutional Review Board Statement: Not applicable.

Informed Consent Statement: Not applicable.

Data Availability Statement: Not applicable.

Conflicts of Interest: The authors declare no conflict of interest.

References

1. Bourgonje, A.R.; Abdulle, A.E.; Timens, W.; Hillebrands, J.L.; Navis, G.J.; Gordijn, S.J.; Bolling, M.C.; Dijkstra, G.; Voors, A.A.; Osterhaus, A.D.; et al. Angiotensin-converting enzyme 2 (ACE2), SARS-CoV-2 and the pathophysiology of coronavirus disease 2019 (COVID-19). *J. Pathol.* **2020**, *251*, 228–248. [CrossRef]
2. Xu, Z.; Shi, L.; Wang, Y.; Zhang, J.; Huang, L.; Zhang, C.; Liu, S.; Zhao, P.; Liu, H.; Zhu, L.; et al. Pathological findings of COVID-19 associated with acute respiratory distress syndrome. *Lancet Respir. Med.* **2020**, *8*, 420–422. [CrossRef]
3. Ioannidis, J.P.A. Infection fatality rate of COVID-19 inferred from seroprevalence data. *Bull. World Health Organ.* **2021**, *99*, 19–33F. [CrossRef]
4. Rai, D.K.; Sharma, P.; Kumar, R. Post COVID-19 pulmonary fibrosis. Is it real threat? *Indian J. Tuberc.* **2021**, *68*, 330–333. [CrossRef] [PubMed]
5. Han, X.; Cao, Y.; Jiang, N.; Chen, Y.; Alwalid, O.; Zhang, X.; Gu, J.; Dai, M.; Liu, J.; Zhu, W.; et al. Novel Coronavirus Disease 2019 (COVID-19) Pneumonia Progression Course in 17 Discharged Patients: Comparison of Clinical and Thin-Section Computed Tomography Features During Recovery. *Clin. Infect. Dis.* **2020**, *71*, 723–731. [CrossRef]
6. Wang, Y.; Dong, C.; Hu, Y.; Li, C.; Ren, Q.; Zhang, X.; Shi, H.; Zhou, M. Temporal Changes of CT Findings in 90 Patients with COVID-19 Pneumonia: A Longitudinal Study. *Radiology* **2020**, *296*, E55–E64. [CrossRef] [PubMed]
7. Liu, D.; Zhang, W.; Pan, F.; Li, L.; Yang, L.; Zheng, D.; Wang, J.; Liang, B. The pulmonary sequelae in discharged patients with COVID-19: A short-term observational study. *Respir. Res.* **2020**, *21*, 125. [CrossRef] [PubMed]
8. Carsana, L.; Sonzogni, A.; Nasr, A.; Rossi, R.S.; Pellegrinelli, A.; Zerbi, P.; Rech, R.; Colombo, R.; Antinori, S.; Corbellino, M.; et al. Pulmonary post-mortem findings in a series of COVID-19 cases from northern Italy: A two-centre descriptive study. *Lancet Infect. Dis.* **2020**, *20*, 1135–1140. [CrossRef] [PubMed]

9. Lechowicz, K.; Drozdzal, S.; Machaj, F.; Rosik, J.; Szosta, B.; Zegan-Baranska, M.; Biernawska, J.; Dabrowski, W.; Rotter, I.; Kotfis, K. COVID-19: The potential treatment of pulmonary fibrosis associated with SARS-CoV-2 infection. *J. Clin. Med.* **2020**, *9*, 1917. [CrossRef] [PubMed]
10. Han, X.; Fan, Y.; Alwalid, O.; Li, N.; Jia, X.; Yuan, M.; Li, Y.; Cao, Y.; Gu, J.; Wu, H.; et al. Six-month Follow-up Chest CT Findings after Severe COVID-19 Pneumonia. *Radiology* **2021**, *299*, E177–E186. [CrossRef]
11. Krishna, R.; Chapman, K.; Ullah, S. *Idiopathic Pulmonary Fibrosis*; StatPearls Publishing: Treasure Island, FL, USA, 2023.
12. Kawabata, Y. Pathology of IPF. In *Idiopathic Pulmonary Fibrosis*; Nakamura, H., Aoshiba, K., Eds.; Springer: Tokyo, Japan, 2016.
13. Barratt, S.L.; Creamer, A.; Hayton, C.; Chaudhuri, N. Idiopathic Pulmonary Fibrosis (IPF): An Overview. *J. Clin. Med.* **2018**, *7*, 201. [CrossRef] [PubMed]
14. Fadista, J.; Kraven, L.M.; Karjalainen, J.; Andrews, S.J.; Geller, F. Shared genetic aetiology between idiopathic pulmonary fibrosis and COVID-19 severity. *EBioMedicine* **2021**, *65*, 103277. [CrossRef] [PubMed]
15. Marvisi, M.; Ferrozzi, F.; Balzarini, L.; Mancini, C.; Ramponi, S.; Uccelli, M. First report on clinical and radiological features of COVID-19 pneumonitis in a Caucasian population: Factors predicting fibrotic evolution. *Int. J. Infect. Dis.* **2020**, *99*, 485–488. [CrossRef]
16. Ojo, A.S.; Balogun, S.A.; Williams, O.T.; Ojo, O.S. Pulmonary Fibrosis in COVID-19 Survivors: Predictive Factors and Risk Reduction Strategies. *Pulm. Med.* **2020**, *2020*, 1–10. [CrossRef]
17. Huang, W.; Wu, Q.; Chen, Z.; Xiong, Z.; Wang, K.; Tian, J.; Zhang, S. The potential indicators for pulmonary fibrosis in survivors of severe COVID-19. *J. Infect.* **2020**, *82*, e5–e7. [CrossRef]
18. Yu, M.; Liu, Y.; Xu, D.; Zhang, R.; Lan, L.; Xu, H. Prediction of the Development of Pulmonary Fibrosis Using Serial Thin-Section CT and Clinical Features in Patients Discharged after Treatment for COVID-19 Pneumonia. *Korean J. Radiol.* **2020**, *21*, 746–755. [CrossRef]
19. Liu, W.; Tao, Z.W.; Wang, L.; Yuan, M.L.; Liu, K.; Zhou, L.; Wei, S.; Deng, Y.; Liu, J.; Liu, H.G.; et al. Analysis of factors associated with disease outcomes in hospitalized patients with 2019 novel coronavirus disease. *Chinese Med. J.* **2020**, *133*, 1032e1038. [CrossRef]
20. Vardavas, C.I.; Nikitara, K. COVID-19 and smoking: A systematic review of the evidence. *Tob. Induc. Dis.* **2020**, *18*, 20. [CrossRef]
21. Jensen, K.; Nizamutdinov, D.; Guerrier, M.; Afroze, S.; Dostal, D.; Glaser, S. General mechanisms of nicotine-induced fibrogenesis. *FASEB J.* **2012**, *26*, 4778–4787. [CrossRef] [PubMed]
22. Camelo, A.; Dunmore, R.; Sleeman, M.A.; Clarke, D.L. The epithelium in idiopathic pulmonary fibrosis: Breaking the barrier. *Front. Pharmacol.* **2014**, *4*, 173. [CrossRef]
23. Vasarmidi, E.; Tsitoura, E.; Spandidos, D.A.; Tzanakis, N.; Antoniou, K.M. Pulmonary fibrosis in the aftermath of the Covid-19 era (Review). *Exp. Ther. Med.* **2020**, *20*, 2557–2560. [CrossRef] [PubMed]
24. Guler, S.A.; Ebner, L.; Beigelman, C.; Bridevaux, P.O.; Brutsche, M.; Clarenbach, C.; Garzoni, C.; Geiser, T.K.; Lenoir, A.; Mancinetti, M.; et al. Pulmonary function and radiological features four months after COVID-19: First results from the national prospective observational Swiss COVID-19 lung study. *Eur. Respir. J.* **2021**, *57*, 2003690. [CrossRef]
25. McGroder, C.F.; Zhang, D.; Choudhury, M.A.; Salvatore, M.M.; D'Souza, B.M.; Hoffman, E.A.; Wei, Y.; Baldwin, M.R.; Garcia, C.K. Pulmonary fibrosis four months after COVID-19 is associated with severity of illness and blood leucocyte telomere length. *Thorax* **2021**, *76*, 1242–1245. [CrossRef] [PubMed]
26. Rogliani, P.; Calzetta, L.; Coppola, A.; Puxeddu, E.; Sergiacomi, G.; D'Amato, D.; Orlacchio, A. Are there pulmonary sequelae in patients recovering from COVID-19? *Respir. Res.* **2020**, *21*, 286. [CrossRef] [PubMed]
27. Ding, M.; Zhang, Q.; Li, Q.; Wu, T.; Huang, Y.-Z. Correlation analysis of the severity and clinical prognosis of 32 cases of patients with COVID-19. *Respir. Med.* **2020**, *167*, 105981. [CrossRef] [PubMed]
28. Hu, Z.J.; Xu, J.; Yin, J.M.; Li, L.; Hou, W.; Zhang, L.L.; Zhou, Z.; Yu, Y.Z.; Li, H.J.; Feng, Y.M.; et al. Lower Circulating Interferon-Gamma Is a Risk Factor for Lung Fibrosis in COVID-19 Patients. *Front. Immunol.* **2020**, *11*, 585647. [CrossRef]
29. Zaman, T.; Lee, J.S. Risk Factors for the Development of Idiopathic Pulmonary Fibrosis: A Review. *Curr. Pulmonol. Rep.* **2018**, *7*, 118–125. [CrossRef]
30. Reyfman, P.A.; Gottardi, C.J. IPF and lung cancer: Finding similarities within differences. *Am. J. Respir. Cell Mol. Biol.* **2019**, *61*, 667–668. [CrossRef]
31. Álvarez, D.; Cárdenes, N.; Sellarés, J.; Bueno, M.; Corey, C.; Hanumanthu, V.S.; Peng, Y.; D'cunha, H.; Sembrat, J.; Nouraie, M.; et al. IPF lung fibroblasts have a senescent phenotype. *Am. J. Physiol. Cell. Mol. Physiol.* **2017**, *313*, L1164–L1173. [CrossRef]
32. Leung, J.; Cho, Y.; Lockey, R.F.; Kolliputi, N. The Role of Aging in Idiopathic Pulmonary Fibrosis. *Lung* **2015**, *193*, 605–610. [CrossRef]
33. Jo, H.E.; Glaspole, I.; Grainge, C.; Goh, N.; Hopkins, P.M.; Moodley, Y.; Reynolds, P.N.; Chapman, S.; Walters, E.H.; Zappala, C.; et al. Baseline characteristics of idiopathic pulmonary fibrosis: Analysis from the Australian Idiopathic Pulmonary Fibrosis Registry. *Eur. Respir. J.* **2017**, *49*, 1601592. [CrossRef]
34. Ali, R.M.M.; Ghonimy, M.B.I. Post-COVID-19 pneumonia lung fibrosis: A worrisome sequelae in surviving patients. *Egypt. J. Radiol. Nucl. Med.* **2021**, *52*, 1–8. [CrossRef]
35. Voltz, J.W.; Card, J.W.; Carey, M.A.; DeGraff, L.M.; Ferguson, C.D.; Flake, G.P.; Bonner, J.C.; Korach, K.S.; Zeldin, D.C. Male Sex Hormones Exacerbate Lung Function Impairment after Bleomycin-Induced Pulmonary Fibrosis. *Am. J. Respir. Cell Mol. Biol.* **2008**, *39*, 45–52. [CrossRef] [PubMed]

36. Lekgabe, E.D.; Royce, S.G.; Hewitson, T.D.; Tang, M.L.K.; Zhao, C.; Moore, X.L.; Tregear, G.W.; Bathgate, R.A.D.; Du, X.-J.; Samuel, C.S. The Effects of Relaxin and Estrogen Deficiency on Collagen Deposition and Hypertrophy of Nonreproductive Organs. *Endocrinology* **2006**, *147*, 5575–5583. [CrossRef]
37. Baumgartner, K.B.; Samet, J.M.; Stidley, C.A.; Colby, T.V.; Waldron, J.A. Cigarette smoking: A risk factor for idiopathic pul-monary fibrosis. *Am. J. Respir. Crit. Care Med.* **1997**, *155*, 242–248. [CrossRef] [PubMed]
38. Taskar, V.S.; Coultas, D.B. Is idiopathic pulmonary fibrosis an environmental disease? *Proc. Am. Thorac. Soc.* **2006**, *3*, 293–298. [CrossRef] [PubMed]
39. Velavan, T.P.; Pallerla, S.R.; Rüter, J.; Augustin, Y.; Kremsner, P.G.; Krishna, S.; Meyer, C.G. Host genetic factors determining COVID-19 susceptibility and severity. *Ebiomedicine* **2021**, *72*, 103629. [CrossRef]
40. Ellinghaus, D.; Degenhardt, F.; Bujanda, L.; Buti, M.; Albillos, A.; Invernizzi, P.; Fernández, J.; Prati, D.; Baselli, G.; Asselta, R.; et al. Severe COVID-19 GWAS Group. Genome-wide Association Study of Severe COVID-19 with Respiratory Failure. *N. Engl. J. Med.* **2020**, *383*, 1522–1534.
41. COVID-19 Host Genetics Initiative. COVID-19 Host Genetics Initiative Mapping the human genetic architecture of COVID-19. *Nature* **2021**, *600*, 472–477. [CrossRef]
42. Wein, A.N.; McMaster, S.R.; Takamura, S.; Dunbar, P.R.; Cartwright, E.K.; Hayward, S.L.; McManus, D.T.; Shimaoka, T.; Ueha, S.; Tsukui, T.; et al. CXCR6 regulates localisation of tissue-resident memory CD8 T cells to the airways. *J. Exp. Med.* **2019**, *216*, 2748–2762. [CrossRef]
43. Kropski, J.A.; Blackwell, T.S.; Loyd, J.E. The genetic basis of idiopathic pulmonary fibrosis. *Eur. Respir. J.* **2015**, *45*, 1717–1727. [CrossRef]
44. Kaur, A.; Mathai, S.K.; Schwartz, D.A. Genetics in Idiopathic Pulmonary Fibrosis Pathogenesis, Prognosis, and Treatment. *Front. Med.* **2017**, *4*, 154. [CrossRef]
45. van Moorsel, C.H.M.; van der Vis, J.J.; Benschop, C.; Ruven, H.J.T.; Quanjel, M.; Grutters, J.C. THE MUC5B promotor polymorphism associates with severe COVID-19. *Front. Med.* **2021**, *8*, 668024. [CrossRef]
46. Molyneaux, P.L.; Cox, M.; Willis-Owen, S.A.; Mallia, P.; Russell, K.E.; Russell, A.M.; Murphy, E.; Johnston, S.L.; Schwartz, D.A.; Wells, A.U.; et al. The role of bacteria in the pathogenesis and progression of idiopathic pulmonary fibrosis. *Am. J. Respir. Crit. Care Med.* **2014**, *190*, 906–913. [CrossRef]
47. Han, M.K.; Zhou, Y.; Murray, S.; Tayob, N.; Noth, I.; Lama, V.N.; Moore, B.B.; White, E.S.; Flaherty, K.R.; Huffnagle, G.B.; et al. Lung microbiome and disease progression in idiopathic pulmonary fibrosis: An analysis of the COMET study. *Lancet Respir. Med.* **2014**, *2*, 548–556. [CrossRef] [PubMed]
48. Nasir, N.; Rehman, F.; Omair, S.F. Risk factors for bacterial infections in patients with moderate to severe COVID-19: A case-control study. *J. Med. Virol.* **2021**, *93*, 4564–4569. [CrossRef] [PubMed]
49. Langford, B.J.; So, M.; Raybardhan, S.; Leung, V.; Westwood, D.; MacFadden, D.R.; Soucy, J.R.; Daneman, N. Bacterial co-infection and secondary infection in patients with COVID-19: A living rapid review and meta-analysis. *Clin. Microbiol. Infect.* **2020**, *26*, 1622–1629. [CrossRef] [PubMed]
50. Ahmad, A.M.; Ata, F.; Islam, A.M.; Bint, I.B.A.; Salih, A.A.; Yousaf, Z. Post COVID-19 fibrosis, an emerging complication of SARS-CoV-2 infection. *IDCases* **2020**, *23*, e01041. [CrossRef] [PubMed]
51. Lynch, D.A.; Sverzellati, N.; Travis, W.D.; Brown, K.K.; Colby, T.V.; Galvin, J.R.; Goldin, J.G.; Hansell, D.M.; Inoue, Y.; Johkoh, T.; et al. Diagnostic criteria for idiopathic pulmonary fibrosis: A Fleischner Society White Paper. *Lancet Respir. Med.* **2018**, *6*, 138–153. [CrossRef] [PubMed]
52. Farghaly, S.; Badedi, M.; Ibrahim, R.; Sadhan, M.H.; Alamoudi, A.; Alnami, A.; Muhajir, A. Clinical characteristics and out-comes of post-COVID-19 pulmonary fibrosis: A case-control study. *Medicine* **2022**, *101*, e28639. [CrossRef]
53. Kamal, M.; Abo Omirah, M.; Hussein, A.; Saeed, H. Assessment and characterisation of post-COVID-19 manifestations. *Int. J. Clin. Pract.* **2021**, *75*, e13746. [CrossRef]
54. Nolan, C.M.; Patel, S.; Barker, R.E.; George, P.; Maddocks, M.M.; Cullinan, P.; Maher, T.M.; Man, W.D.C. Anxiety and depression in idiopathic pulmonary fibrosis (IPF): Prevalence and clinical correlates. *Eur. Respir. J.* **2017**, *50*, PA848.
55. Lechtzin, N.; Hilliard, M.E.; Horton, M.R. Validation of the Cough Quality-of-Life Questionnaire in Patients with Idiopathic Pulmonary Fibrosis. *Chest* **2013**, *143*, 1745–1749. [CrossRef]
56. Araki, T.; Katsura, H.; Sawabe, M.; Kida, K. A Clinical Study of Idiopathic Pulmonary Fibrosis Based on Autopsy Studies in Elderly Patients. *Intern. Med.* **2003**, *42*, 483–489. [CrossRef]
57. Carvajalino, S.; Reigada, C.; Johnson, M.J.; Dzingina, M.; Bajwah, S. Symptom prevalence of patients with fibrotic interstitial lung disease: A systematic literature review. *BMC Pulm. Med.* **2018**, *18*, 78. [CrossRef] [PubMed]
58. Faverio, P.; Luppi, F.; Rebora, P.; Busnelli, S.; Stainer, A.; Catalano, M.; Parachini, L.; Monzani, A.; Galimberti, S.; Bini, F.; et al. Six-Month Pulmonary Impairment after Severe COVID-19: A Prospective, Multicentre Follow-Up Study. *Respiration* **2021**, *100*, 1078–1087. [CrossRef] [PubMed]
59. Kanematsu, T.; Kitaichi, M.; Nishimura, K.; Nagai, S.; Izumi, T. Clubbing of the Fingers and Smooth-Muscle Proliferation in Fibrotic Changes in the Lung in Patients With Idiopathic Pulmonary Fibrosis. *Chest* **1994**, *105*, 339–342. [CrossRef]
60. Bonella, F.; di Marco, F.; Spagnolo, P. Pulmonary Function Tests in Idiopathic Pulmonary Fibrosis. In *Idiopathic Pulmonary Fibrosis. Respiratory Medicine*; Meyer, K., Nathan, S., Eds.; Humana Press: Cham, Switzerland, 2019.

61. Torres, C.R.; Vasconcello, C.L.; Alsina, R.L.; Solis, N.F.; Burgos, H.; Puppo, J.V. Respiratory function in patients post-infection by COVID-19: A systematic review and meta-analysis. *Pulmonology* **2021**, *27*, 328–337. [CrossRef]
62. British Thoracic Society. British Thoracic Society Guidance on Respiratory Follow-Up of Patients with a Clinico-Radiological Diagnosis of COVID-19 Pneumonia [Internet]. 2020. Available online: https://www.brit-thoracic.org.uk/document-library/quality-improvement/COVID-19/resp-follow-up-guidance-post-COVID-pneumonia/ (accessed on 2 July 2022).
63. Cherrez-Ojeda, I.; Robles-Velasco, K.; Osorio, M.F.; Cottin, V.; Centeno, J.V.; Felix, M. Follow-up of two cases of suspected interstitial lung disease following severe COVID-19 infection shows persistent changes in imaging and lung function. *Clin. Case Rep.* **2021**, *9*, e04918. [CrossRef] [PubMed]
64. Cabo-Gambin, R.; Benítez, I.D.; Carmona, P.; Santiesteve, S.; Mínguez, O.; Vaca, R.; Moncusí-Moix, A.; Gort-Paniello, C.; García-Hidalgo, M.C.; de Gonzalo-Calvo, D.; et al. Three to Six Months Evolution of Pulmonary Function and Radiological Features in Critical COVID-19 Patients: A Prospective Cohort. *Arch. Bronconeumol.* **2021**, *58*, 59–62. [CrossRef]
65. Fidler, L.; Shapera, S.; Mittoo, S.; Marras, T.K. Diagnostic disparity of previous and revised American Thoracic Society guide-lines for idiopathic pulmonary fibrosis. *Can. Respir. J.* **2015**, *22*, 86–90. [CrossRef]
66. Collard, H.R.; King, T.E.; Bartelson, B.B.; Vourlekis, J.S.; Schwarz, M.I.; Brown, K.K. Changes in clinical and physiologic var-iables predict survival in idiopathic pulmonary fibrosis. *Am. J. Respir. Crit. Care Med.* **2003**, *168*, 538–542. [CrossRef]
67. Nathan, S.D.; Shlobin, O.A.; Weir, N.; Ahmad, S.; Kaldjob, J.M.; Battle, E.; Sheridan, M.J.; du Bois, R.M. Long-term course and prognosis of idiopathic pulmonary fibrosis in the new millennium. *Chest* **2011**, *140*, 221–229. [CrossRef]
68. Tanni, S.E.; Fabro, A.T.; de Albuquerque, A.; Ferreira, E.V.M.; Verrastro, C.G.Y.; Sawamura, M.V.Y.; Ribeiro, S.M.; Baldi, B.G. Pulmonary fibrosis secondary to COVID-19: A narrative review. *Exp. Rev. Respir. Med.* **2021**, *15*, 791–803. [CrossRef]
69. Atabati, E.; Dehghani-Samani, A.; Mortazavimoghaddam, S.G. Association of COVID-19 and other viral infections with interstitial lung diseases, pulmonary fibrosis, and pulmonary hypertension: A narrative review. *Can. J. Respir. Ther.* **2020**, *56*, 70–78. [CrossRef] [PubMed]
70. Wie, J.; Yang, H.; Lei, P.; Fan, B.; Qiu, Y.; Zeng, B. Analysis of thin-section CT in patients with coronavirus disease (COVID-19) after hospital discharge. *J. X-ray Sci. Technol.* **2020**, *28*, 383–389.
71. Francone, M.; Iafrate, F.; Masci, G.M.; Coco, S.; Cilia, F.; Manganaro, L.; Panebianco, V.; Andreoli, C.; Colaiacomo, M.C.; Zingaropoli, M.A.; et al. Chest CT score in COVID-19 patients: Correlation with disease severity and short-term prognosis. *Eur. Radiol.* **2020**, *30*, 6808–6817. [CrossRef] [PubMed]
72. Yin, X.; Xi, X.; Min, X.; Feng, Z.; Li, B.; Cai, W.; Fan, C.; Wang, L.; Xia, L. Long-term chest CT follow-up in COVID-19 Survivors: 102–361 days after onset. *Ann. Transl. Med.* **2021**, *9*, 1231. [CrossRef]
73. Vijayakumar, B.; Tonkin, J.; Devaraj, A.; Philip, K.E.J.; Orton, C.M.; Desai, S.R.; Shah, P.L. CT Lung Abnormalities after COVID-19 at 3 Months and 1 Year after Hospital Discharge. *Radiology* **2022**, *303*, 444–454. [CrossRef]
74. Camiciottoli, G.; Orlandi, I.; Bartolucci, M.; Meoni, E.; Nacci, F.; Diciotti, S.; Barcaroli, C.; Conforti, M.L.; Pistolesi, M.; Matucci-Cerinic, M.; et al. Lung CT densitometry in systemic sclerosis: Correlation with lung function, exercise testing, and quality of life. *Chest* **2007**, *131*, 672–681. [CrossRef]
75. Combet, M.; Pavot, A.; Savale, L.; Humbert, M.; Monnet, X. Rapid onset honeycombing fibrosis in spontaneously breathing patient with COVID-19. *Eur. Respir. J.* **2020**, *56*, 2001808. [CrossRef] [PubMed]
76. Balestro, E.; Cocconcelli, E.; Giraudo, C.; Polverosi, R.; Biondini, D.; Lacedonia, D.; Bazzan, E.; Mazzai, L.; Rizzon, G.; Lococo, S.; et al. High-Resolution CT Change over Time in Patients with Idiopathic Pulmonary Fibrosis on Anti-fibrotic Treatment. *J. Clin. Med.* **2019**, *8*, 1469. [CrossRef] [PubMed]
77. Ley, B.; Elicker, B.M.; Hartman, T.E.; Ryerson, C.J.; Vittinghoff, E.; Ryu, J.H.; Lee, J.S.; Jones, K.D.; Richeldi, L.; King, T.E.; et al. Idiopathic Pulmonary Fibrosis: CT and Risk of Death. *Radiology* **2014**, *273*, 570–579. [CrossRef]
78. Raghu, G.; Collard, H.R.; Egan, J.J.; Martinez, F.J.; Behr, J.; Brown, K.K.; Colby, T.V.; Cordier, J.F.; Flaherty, K.R.; Lasky, J.A.; et al. An Official ATS/ERS/JRS/ALAT Statement: Idiopathic Pulmonary Fibrosis: Evidence-based Guidelines for Diagnosis and Management. *Am. J. Respir. Crit. Care Med.* **2011**, *183*, 788–824. [CrossRef]
79. Adachi, T.; Chong, J.M.; Nakajima, N.; Sano, M.; Yamazaki, J.; Miyamoto, I.; Nishioka, H.; Akita, H.; Sato, Y.; Kataoka, M.; et al. Clinicopathologic and Immunohistochemical Findings from Autopsy of Patient with COVID-19, Japan. *Emerg. Infect. Dis.* **2020**, *26*, 2157–2161. [CrossRef] [PubMed]
80. Vishwajit, D.; Rohini, M.; Ashutosh, K.; Chiman, K.; Khursheed, R. Histopathological observations in COVID-19: A systematic review. *J. Clin. Pathol.* **2021**, *74*, 76–83.
81. Martines, R.B.; Ritter, J.M.; Matkovic, E.; Gary, J.; Bollweg, B.C.; Bullock, H.; Goldsmith, C.S.; Silva-Flannery, L.; Seixas, J.N.; Reagan-Steiner, S.; et al. COVID-19 pathology working group. Pathology and pathogenesis of SARS-CoV-2 associated with fatal coronavirus disease, United States. *Emerg. Infect. Dis.* **2020**, *26*, 2005–2015. [CrossRef]
82. Sauter, J.L.; Baine, M.K.; Butnor, K.J.; Buonocore, D.J.; Chang, J.C.; Jungbluth, A.A.; Szabolcs, M.J.; Morjaria, S.; Mount, S.L.; Rekhtman, N.; et al. Insights into pathogenesis of fatal COVID-19 pneumonia from histopathology with immunohistochemical and viral RNA studies. *Histopathology* **2020**, *77*, 915–925. [CrossRef]
83. Chen, J.Y.; Qiao, K.; Liu, F.; Wu, B.; Xu, X.; Jiao, G.Q.; Lu, R.G.; Li, H.X.; Zhao, J.; Huang, J.; et al. Lung transplantation as therapeutic option in acute respiratory distress syndrome for coronavirus disease 2019-related pulmonary fibrosis. *Chin. Med. J.* **2020**, *133*, 1390–1396. [CrossRef]

84. Tian, S.; Xiong, Y.; Liu, H.; Niu, L.; Guo, J.; Liao, M.; Xiao, S.Y. Pathological study of the 2019 novel coronavirus disease (COVID-19) through post-mortem core biopsies. *Mod. Pathol.* **2020**, *33*, 1007–1014. [CrossRef]
85. Bernard, G.R.; Artigas, A.; Brigham, K.L.; Carlet, J.; Falke, K.; Hudson, L.; Lamy, M.; Legall, J.R.; Morris, A.; Spragg, R. The American-European Consensus Conference on ARDS. Definitions, mechanisms, relevant outcomes, and clinical trial coordination. *Am. J. Respir. Crit. Care Med.* **1994**, *149*, 818–824. [CrossRef] [PubMed]
86. Beasley, M.B.; Franks, T.J.; Galvin, J.R.; Gochuico, B.; Travis, W.D. Acute fibrinous and organising pneumonia: A histologic pattern of lung injury and possible variant of diffuse alveolar damage. *Arch. Pathol. Lab. Med.* **2002**, *126*, 1064–1070. [CrossRef]
87. Tomashefski, J.F., Jr. Pulmonary Pathology of Acute Respiratory Distress Syndrome. *Clin. Chest Med.* **2000**, *21*, 435–466. [CrossRef]
88. Sapru, A.; Wiemels, J.L.; Witte, J.S.; Ware, L.B.; Matthay, M.A. Acute lung injury and the coagulation pathway: Potential role of gene polymorphisms in the protein C and fibrinolytic pathways. *Intensiv. Care Med.* **2006**, *32*, 1293–1303. [CrossRef] [PubMed]
89. Li, Y.; Wu, J.; Wang, S.; Li, X.; Zhou, J.; Huang, B.; Luo, D.; Cao, Q.; Chen, Y.; Chen, S.; et al. Progression to fibrosing diffuse alveolar damage in a series of 30 minimally invasive autopsies with COVID-19 pneumonia in Wuhan, China. *Histopathology* **2020**, *78*, 542–555. [CrossRef]
90. Zhang, H.; Zhou, P.; Wei, Y.; Yue, H.; Wang, Y.; Hu, M.; Zhang, S.; Cao, T.; Yang, C.; Li, M.; et al. Histopathologic Changes and SARS-CoV-2 immunostaining in the lung of a patient with COVID-19. *Ann. Intern. Med.* **2020**, *172*, 629–632. [CrossRef] [PubMed]
91. Kory, P.; Jp, K. SARS-CoV-2 organising pneumonia: Has there been a widespread failure to identify and treat this prevalent condition in COVID-19? *BMJ Open Respir. Res.* **2020**, *7*, e000724. [CrossRef]
92. Kommoss, F.K.; Schwab, C.; Tavernar, L.; Schreck, J.; Wagner, W.L.; Merle, U.; Jonigk, D.; Schirmacher, P.; Longerich, T. The Pathology of Severe COVID-19-Related Lung Damage. *Deutsches Ärzteblatt Int.* **2020**, *117*, 500–506. [CrossRef]
93. Hanley, B.; Naresh, K.N.; Roufosse, C.; Nicholson, A.G.; Weir, J.; Cooke, G.S.; Thursz, M.; Manousou, P.; Corbett, R.; Goldin, R.; et al. Histopathological findings and viral tropism in UK patients with severe fatal COVID-19: A post-mortem study. *Lancet Microbe* **2020**, *1*, e245–e253. [CrossRef]
94. Copin, M.C.; Parmentier, E.; Duburcq, T.; Poissy, J.; Mathieu, D.; The Lille COVID-19 ICU and Anatomopathology Group. Time to consider histologic pattern of lung injury to treat critically ill patients with COVID-19 infection. *Intensiv. Care Med.* **2020**, *46*, 1124–1126. [CrossRef]
95. Bharat, A.; Querrey, M.; Markov, N.S.; Kim, S.; Kurihara, C.; Garza-Castillon, R.; Manerikar, A.; Shilatifard, A.; Tomic, R.; Politanska, Y.; et al. Lung transplantation for patients with severe COVID-19. *Sci. Transl. Med.* **2020**, *12*. [CrossRef]
96. John, A.E.; Joseph, C.; Jenkins, G.; Tatler, A.L. COVID-19 and pulmonary fibrosis: A potential role for lung epithelial cells and fibroblasts. *Immunol. Rev.* **2021**, *302*, 228–240. [CrossRef]
97. Buja, L.M.; Wolf, D.A.; Zhao, B.; Akkanti, B.; McDonald, M.; Lelenwa, L.; Reilly, N.; Ottaviani, G.; Elghetany, M.T.; Trujillo, D.O.; et al. The emerging spectrum of cardiopulmonary pathology of the coronavirus disease 2019 (COVID-19): Report of 3 autopsies from Houston, Texas, and review of autopsy findings from other United States cities. *Cardiovasc. Pathol.* **2020**, *48*, 107233. [CrossRef]
98. Ackermann, M.; Verleden, S.E.; Kuehnel, M. Pulmonary vascular endothelialitis, thrombosis, and angiogenesis in COVID-19. *N Engl. J. Med.* **2020**, *383*, 120–128. [CrossRef]
99. Burel-Vandenbos, F.; Cardot-Leccia, N.; Passeron, T. Pulmonary Vascular Pathology in Covid-19. *New Engl. J. Med.* **2020**, *383*, 886–889. [PubMed]
100. Varga, Z.; Flammer, A.J.; Steiger, P.; Haberecker, M.; Andermatt, R.; Zinkernagel, A.S.; Mehra, M.R.; Schuepbach, R.A.; Ruschitzka, F.; Moch, H. Endothelial cell infection and endotheliitis in COVID-19. *Lancet* **2020**, *395*, 1417–1418. [CrossRef] [PubMed]
101. Nienhold, R.; Ciani, Y.; Koelzer, V.H.; Tzankov, A.; Haslbauer, J.D.; Menter, T.; Schwab, N.; Henkel, M.; Frank, A.; Zsikla, V.; et al. Two distinct immunopathological profiles in autopsy lungs of COVID-19. *Nat. Commun.* **2020**, *11*, 1–13. [CrossRef]
102. Raghu, G.; Remy-Jardin, M.; Myers, J.L.; Richeldi, L.; Ryerson, C.J.; Lederer, D.J.; Behr, J.; Cottin, V.; Danoff, S.K.; Morell, F.; et al. Diagnosis of Idiopathic Pulmonary Fibrosis. An Official ATS/ERS/JRS/ALAT Clinical Practice Guideline. *Am. J. Respir. Crit. Care Med.* **2018**, *198*, e44–e68. [CrossRef]
103. Parambil, J.G.; Myers, J.L.; Aubry, M.C.; Ryu, J.H. Causes and Prognosis of Diffuse Alveolar Damage Diagnosed on Surgical Lung Biopsy. *Chest* **2007**, *132*, 50–57. [CrossRef] [PubMed]
104. Fisher, M.; Nathan, S.D.; Hill, C.; Marshall, J.; Dejonckheere, F.; Thuresson, P.O.; Maher, T.M. Predicting Life Expectancy for Pirfenidone in Idiopathic Pulmonary Fibrosis. *J. Manag. Care Spéc. Pharm.* **2017**, *23*, S17–S24. [CrossRef] [PubMed]
105. Guan, W.J.; Ni, Z.Y.; Hu, Y. Clinical characteristics of coronavirus disease 2019 in China. *N. Engl. J. Med.* **2020**, *382*, 1708–1720. [CrossRef] [PubMed]
106. Tang, N.; Bai, H.; Chen, X.; Gong, J.; Li, D.; Sun, Z. Anticoagulant treatment is associated with decreased mortality in severe coronavirus disease 2019 patients with coagulopathy. *J. Thromb. Haemost.* **2020**, *18*, 1094–1099. [CrossRef] [PubMed]
107. Jolly, L.; Stavrou, A.; Vanderstoken, G. Influenza promotes collagen deposition via αvβ6 integrin-mediated transforming growth factor β activation. *J. Biol. Chem.* **2014**, *289*, 35246–35263. [CrossRef]
108. Mackinnon, A.C.; Gibbons, M.A.; Farnworth, S.L. Regulation of transforming growth factor-β1-driven lung fibrosis by galec-tin-3. *Am. J. Respir. Crit. Care Med.* **2012**, *185*, 537–546. [CrossRef]
109. Gordon, D.E.; Jang, G.M.; Bouhaddou, M.A. SARS-CoV-2 protein interaction map reveals targets for drug repurposing. *Nature* **2020**, *583*, 459–468. [CrossRef]

110. Raghu, G.; van den Blink, B.; Hamblin, M.J. Effect of recombinant human pentraxin 2 vs placebo on change in forced vital capacity in patients with idiopathic pulmonary fibrosis: A randomised clinical trial. *JAMA* **2018**, *319*, 2299–2307. [CrossRef]
111. Liu, Y.; Huang, F.; Xu, J. Anti-hypertensive Angiotensin II receptor blockers associated to mitigation of disease severity in elderly COVID-19 patients. *Ann. Intern. Med.* **2020**, *172*, 629–632.
112. Skurikhin, E.; Nebolsin, V.; Widera, D.; Ermakova, N.; Pershina, O.; Pakhomova, A.; Krupin, V.; Pan, E.; Zhukova, M.; Novikov, F.; et al. Anti-fibrotic and regenerative effects of treamid in pulmonary fibrosis. *Int. J. Mol. Sci.* **2020**, *21*, 8380. [CrossRef]
113. Pure Tech Company. *A Phase 2 Randomised, Double-Blind, Placebo-Controlled Trial and Open Label Extension to Evaluate the Safety and Efficacy of Deupirfenidone (LYT-100) in Post-Acute COVID-19 Respiratory Disease*; Pure Tech Company: Boston, MA, USA, 2020.
114. Bazdyrev, E.; Rusina, P.; Panova, M.; Novikov, F.; Grishagin, I.; Nebolsin, V. Lung Fibrosis after COVID-19: Treatment Pro-spects. *Pharmaceuticals* **2021**, *14*, 807. [CrossRef] [PubMed]
115. Echeverría-Esnal, D.; Martin-Ontiyuelo, C.; Navarrete-Rouco, M.E.; Cuscó, M.D.-A.; Ferrández, O.; Horcajada, J.P.; Grau, S. Azithromycin in the treatment of COVID-19: A review. *Expert Rev. Anti-Infective Ther.* **2020**, *19*, 147–163. [CrossRef] [PubMed]
116. P, K.M.; Sivashanmugam, K.; Kandasamy, M.; Subbiah, R.; Ravikumar, V. Repurposing of histone deacetylase inhibitors: A promising strategy to combat pulmonary fibrosis promoted by TGF-β signalling in COVID-19 survivors. *Life Sci.* **2020**, *266*, 118883. [CrossRef] [PubMed]
117. Reina-Gutiérrez, S.; Torres-Costoso, A.; Martínez-Vizcaíno, V.; de Arenas-Arroyo, S.N.; Fernández-Rodríguez, R.; Pozuelo-Carrascosa, D.P. Effectiveness of Pulmonary Rehabilitation in Interstitial Lung Disease, Including Coronavirus Diseases: A Systematic Review and Meta-analysis. *Arch. Phys. Med. Rehabilit.* **2021**, *102*, 1989–1997.e3. [CrossRef] [PubMed]
118. Egan, J.J. Follow-up and nonpharmacological management of the idiopathic pulmonary fibrosis patient. *Eur. Respir. Rev.* **2011**, *20*, 114–117. [CrossRef] [PubMed]

biomedicines

MDPI

Review

Diagnosis and Treatment of Sleep Apnea in Children: A Future Perspective Is Needed

Esther Solano-Pérez [1,2], Carlota Coso [1,2], María Castillo-García [1,3,4], Sofía Romero-Peralta [1,3,4], Sonia Lopez-Monzoni [1], Eduardo Laviña [1], Irene Cano-Pumarega [2,5], Manuel Sánchez-de-la-Torre [2,6,7], Francisco García-Río [2,8,9,10] and Olga Mediano [1,2,3,*]

1 Sleep Unit, Pneumology Department, Hospital Universitario de Guadalajara, 19002 Guadalajara, Spain
2 Centro de Investigación Biomédica en Red de Enfermedades Respiratorias (CIBERES), 28029 Madrid, Spain
3 Medicine Department, Universidad de Alcalá, 28805 Madrid, Spain
4 Sleep Research Institute, 28036 Madrid, Spain
5 Sleep Unit, Pneumology Department, Hospital Universitario Ramón y Cajal, Instituto Ramón y Cajal de Investigación Sanitaria (IRYCIS), 28034 Madrid, Spain
6 Group of Precision Medicine in Chronic Diseases, University Hospital Arnau de Vilanova and Santa María, 25198 Lleida, Spain
7 Department of Nursing and Physiotherapy, Faculty of Nursing and Physiotherapy, Institut de Recerca Biomèdica de Lleida (IRBLleida), University of Lleida, 25002 Lleida, Spain
8 Respiratory Diseases Group, Respiratory Service, Hospital Universitario La Paz, IdiPAZ, 28046 Madrid, Spain
9 Pneumology Department, Hospital Universitario La Paz, IdiPAZ, 28046 Madrid, Spain
10 Faculty of Medicine, Universidad Autónoma de Madrid, 28049 Madrid, Spain
* Correspondence: olgamediano@hotmail.com; Tel.: +34-949-209-200 (ext. 69522)

Abstract: Obstructive sleep apnea (OSA) in children is a prevalent, but still, today, underdiagnosed illness, which consists of repetitive episodes of upper airway obstruction during sleep with important repercussions for sleep quality. OSA has relevant consequences in the pediatric population, mainly in the metabolic, cardiovascular (CV), and neurological spheres. However, contrary to adults, advances in diagnostic and therapeutic management have been scarce in the last few years despite the increasing scientific evidence of the deleterious consequences of pediatric OSA. The problem of underdiagnosis and the lack of response to treatment in some groups make an update to the management of OSA in children necessary. Probably, the heterogeneity of OSA is not well represented by the classical clinical presentation and severity parameters (apnea/hypopnea index (AHI)), and new strategies are required. A specific and consensus definition should be established. Additionally, the role of simplified methods in the diagnosis algorithm should be considered. Finally, the search for new biomarkers for risk stratification is needed in this population. In conclusion, new paradigms based on personalized medicine should be implemented in this population.

Keywords: sleep apnea; cardiovascular; hypoxic burden; children; diagnosis; treatment

1. Introduction

1.1. Definition and Prevalence of Obstructive Sleep Apnea in Children

Sleep-disordered breathing (SDB) occurs as a result of upper airway (UA) dysfunction (snoring and/or increased respiratory effort). It ranges from snoring to obstructive sleep apnea (OSA), depending on the degree of intermittent UA obstruction [1], and around 20% of children who snore have OSA [2].

OSA is characterized by recurrent events of partial (hypopnea) or complete (apnea) obstructions in the UA, which disrupt normal oxygenation, ventilation, and sleep patterns [1,3–5]. OSA in children has a clear entity with profiles that are very different from adults in terms of etiology, clinical presentation, and consequences (Figure 1). For this reason, a specific definition, diagnosis, and treatment approach is needed for this specific population.

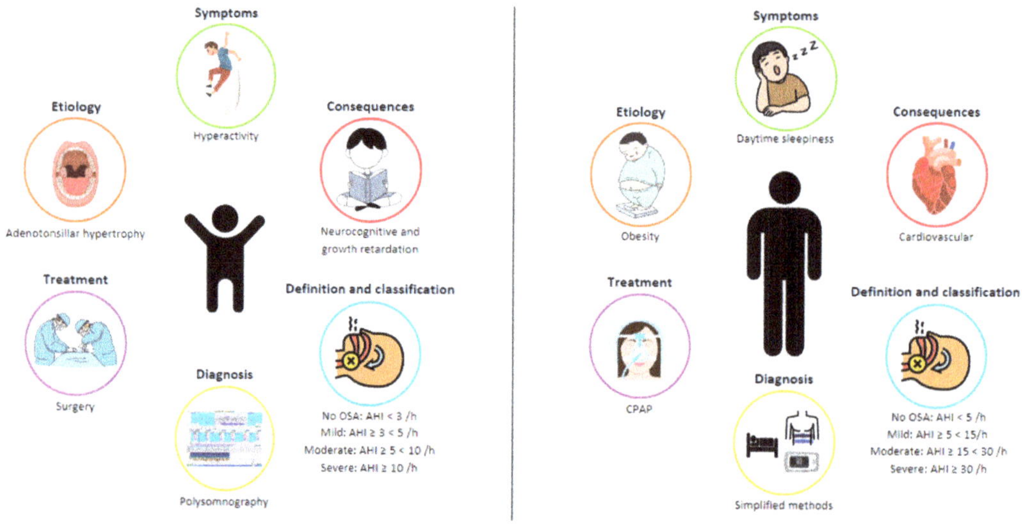

Figure 1. Differences in obstructive sleep apnea (OSA) between adults and children.

OSA is a very frequent condition in children, with prevalence varying between 1 and 4% [6]. Although there has been an effort to increase knowledge about this entity in childhood, there is less scientific evidence than in adults. Different guidelines establish the definition of SDB and OSA in children, although the criteria are diverse and lack recent updates. The classification of OSA severity in children, through the apnea/hypopnea index-(AHI) (number of respiratory events per hour of sleep) obtained from sleep studies, is the most-commonly used parameter. Generally, an AHI of 1–3/h is accepted as the normal cutoff line for the diagnosis of OSA and is classified as follows: mild OSA if the AHI < 5/h, moderate OSA when the AHI is between 5 and 10/h, and severe OSA when the AHI > 10/h [1,3,5,7]. However, these criteria can vary depending on the guidelines, considering factors such as age, additional comorbidities, and other polysomnographic variables (presence and length of oxygen desaturations, degree of hypoventilation, sleep fragmentation, and decreased total sleep time) [7,8].

1.2. Etiology of OSA in Children

The etiology of childhood OSA is multifactorial, involving many risk factors, which can increase UA narrowing and collapsibility and which may contribute to the pathogenesis of OSA [8–10]. This includes both anatomical and neuromuscular disturbances, leading to increased airway resistance and preventing the normal function of the dilator muscles, respectively [5] (Figure 2).

The most-common risk factor is adenotonsillar hypertrophy, reaching the peak of development between 2 and 8 years [11], coinciding with the onset of OSA [12]. Nevertheless, some studies have shown a weak or no correlation between the size of the tonsils and adenoids and the severity of pediatric OSA [13,14]. Craniofacial abnormalities can also be a cause of UA narrowing: alterations of the size, position, and geometry of the mandible and the tongue [10]. These anatomical features are often found in children with craniofacial syndromes, achondroplasia, trisomy 21, Beckwith–Wiedemann syndrome, Chiari malformation, and mucopolysaccharidoses [8].

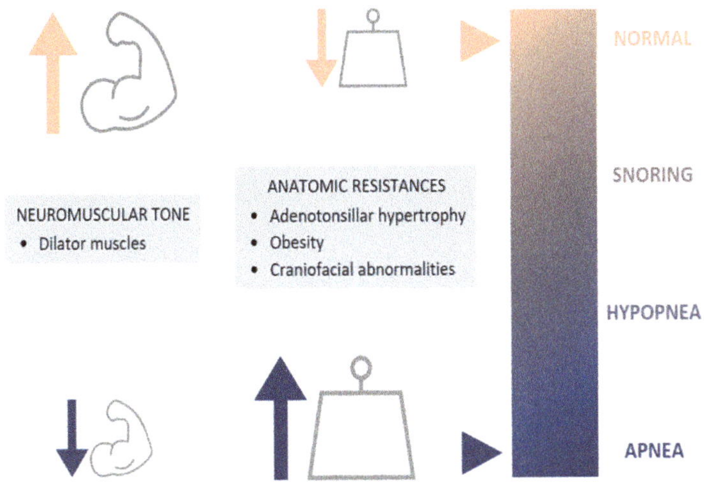

Figure 2. Etiology of pediatric OSA.

Besides anatomic factors, obesity has also been suggested as a contributor to OSA. Obese children represent a special risk factor, as the prevalence of childhood obesity is progressively increasing (5.6% in girls and 7.8% in boys) [15], also leading to an increase in the prevalence of obesity-associated morbidities including OSA [10,11,16]. This relationship is bidirectional, as OSA is known to worsen weight loss and overweight [17,18].

1.3. Symptoms of OSA in Children

The symptoms are classically divided into nocturnal and diurnal (Table 1). Nocturnal symptoms include snoring, witnessed apneas, gasping, oral breathing, paradoxical thoracic movements, nightmares, restless sleep, and nocturnal enuresis. Snoring is the most-common symptom, along with oral breathing. This population can also present disturbed sleep with frequent changes of position, unusual sleep positions (neck hyperextension), and nightmares [9,19]. Enuresis is another frequent symptom in OSA children related to an altered arousal response and sleep fragmentation, often being resolved when OSA is adequately treated [20].

Table 1. Symptoms of OSA in children.

Nocturnal Symptoms	Daytime Symptoms
Snoring	Behavioral disorders
Witnessed apneas	Neurocognitive disorders
Gasping	Mood instability
Oral breathing	Excessive daytime sleepiness
Paradoxical thoracic movements	
Nightmares	
Restless sleep	
Nocturnal enuresis	

Related to daytime symptoms, a relationship between OSA and behavioral disorders (irritability, aggressiveness, and depression), neurocognitive disorders (difficulty concentrating/learning difficulties and inattention), mood instability, and excessive daytime sleepiness has been demonstrated [5,10,21–23].

1.4. Consequences

OSA in children is associated with a number of adverse morbidities, presented as behavioral and neurocognitive disorders, growth retardation, cardiovascular (CV) diseases, and metabolic consequences, producing a negative impact on quality of life. These consequences are derived from the presence of continuous episodes of hypoxia/resaturation, sleep fragmentation, and/or changes in the intrathoracic pressure. These immediate consequences develop a cascade of intermediate mechanisms, mainly alterations in sympathetic activity, coagulation, inflammation, and oxidative stress (Figure 3).

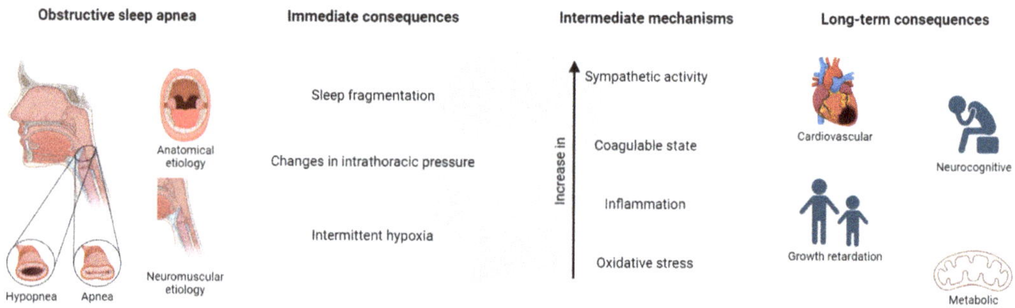

Figure 3. Mechanisms and consequences of OSA in children. Created with BioRender.com. This population is characterized by poor academic performance showing a reduction in memory capacities and difficulties in learning and attention (especially in specific areas such as mathematics, science, reading, and spelling) [24–26], which could be associated with hyperactive behavior during the day.

The effects on growth are probably related to factors such as increased energy consumption and reduced production of growth hormone, whose secretion is characterized by wide and frequent peaks during sleep [9]. These adverse results may be recovered after OSA treatment, as suggested by different studies [27–29].

In the CV sphere, alterations in the autonomic nervous system, vasomotor tone, systemic inflammation, and atherogenesis associated with OSA are likely to induce functional disruption of the endothelium [30]. In addition, many biomarkers have been evaluated to identify this vascular damage, the C-reactive protein (CRP) being the most-studied marker. This inflammatory indicator is increased in children with OSA, with a recent study indicating that it could be reversed after treatment [31]. It has been reported that children with OSA have increased systolic and diastolic blood pressure (BP), increased BP variability, and decreased BP dipping during sleep. Observing the BP of children with OSA is essential to identifying those at risk for developing clinically significant elevated BP in adulthood [32]. There is an independent effect of OSA on cardiopulmonary function, which improves after the disorder is adequately treated [33–35]. Finally, these children may develop an early metabolic syndrome [10,36], this risk being six-times higher than in healthy subjects in adolescents with OSA [37]. A brief literature search of recent evidence in these spheres is described in Table 2.

Table 2. Consequences of OSA in children.

Author (Year)	Number of Participants	Age (Years)	OSA Severity Criteria	Outcomes	Results
			Behavioral and neurocognitive sphere		
Menzies et al., 2022 [38] Metanalysis of 63 studies	17,834	From 2 to 18 years	Due to the lack of a consensus severity criterion, the subgroup given by the author was used (e.g., mild OSA)	Intelligence, attention, memory, visual spatial skills, and language	Children with SDB had significant impairments in all cognitive domains, intelligence being the most-affected quality. These neurocognitive deficits were found in primary snorers among OSA children.
			Growth retardation and metabolism		
Lagravère et al., 2019 [39] Systematic review of 12 studies				Growth mediators (IGF-I and IGFBP-3)	Children with OSA present lower levels of growth mediators, indicating growth retardation, significantly higher cardiovascular disease risk, and decreased cognitive functions compared to healthy controls. Tonsillectomy may improve all these functions with a great impact on general health.
			Cardiovascular sphere		
Ai et al., 2022 [40] Metanalysis of 14 studies	3081	3 to 17 years	Mild OSA is defined as an AHI between 1 and 5 events per hour Moderate to severe is defined as an AHI ≥ 5 /h.	BP parameters: awake and nighttime SBP and DBP	The mean SBP was higher in children with mild or moderate-to-severe OSA compared to healthy controls, these effects being more pronounced during the night. The results suggest that moderate-to-severe OSA in children is associated with a higher risk of adverse SBP outcomes.

Abbreviations: OSA: obstructive sleep apnea; SDB: sleep-disordered breathing; IGF-1: insulin-like growth factor 1; IGFBP3: insulin-like growth factor-binding protein 3; AHI: apnea/hypopnea index; BP: blood pressure; SBP: systolic blood pressure; DBP: diastolic blood pressure.

New evidence regarding untreated pediatric OSA's significant long-term morbidities affecting different organs and systems is necessary [41]. Therefore, in order to minimize the deleterious consequences related to OSA, early correct diagnosis and treatment management are mandatory. The strongest evidence shows that this population may have a significant impact on CV health in childhood and later in adulthood [30]. Tools to identify children at risk and treatment prevention indication are needed.

1.5. Diagnosis

The diagnosis of OSA in the pediatric population differs according to the different clinical guidelines and is described according to the Spanish [5], European [1], and American [3] guidelines in Table 3.

Table 3. Diagnosis and management of OSA in children.

Guide	Diagnosis and Management of OSA in Children
Spanish Society of Pneumology and Thoracic Surgery (SEPAR)	This guide divides the OSA diagnostic methodology between primary care and hospital care in order to increase the diagnostic efficiency. In primary care, the evaluation of the child with suspected OSA (presence of snoring and symptoms or suggestive clinical findings) should include the medical history and complete clinical examination. 1. Medical history: family history, events related to the child's sleep and breathing, and sleep questionnaire (Chervin). 2. Complete clinical examination: craniofacial and UA anatomy, cardiopulmonary examination and somatometry. Children with obesity represent a special risk group. Depending on the results, referral of the patient from primary care to the reference sleep unit is considered. 1. If there is suspected OSA in the clinical history and or/Chervin, retrognathia, adenotonsillar hypertrophy and Mallampati $\geq$ 2, hospital RP is performed. Otherwise, a control visit is carried out 6 months after baseline visit. 2. When the index of respiratory events is $\geq$ 5 in the RP, children are referred to adenotonsillar surgery. With an inferior result, a PSG is performed. 3. An AHI $\geq$ 5/h on PSG leads to adenotonsillar surgery. If the AHI is < 3/h, an anti-inflammatory therapy or review visit after 6 months is assessed. For an AHI 3–5/h comorbidities are evaluated. The presence of comorbidity leads to adenotonsillectomy and anti-inflammatory therapy is selected when there is absence of comorbidity. 4. All children should be clinically reassessed after surgery (recommended in the next 6 months), performing a sleep study in children with severe preoperative OSA or when risk factors or OSA symptoms persist, where other treatments such as diet, CPAP, or orthodontics will be assessed.
European Respiratory Society	The diagnosis and management for SDB is described as a stepwise approach in 7 steps. 1. Identification of risk of SDB: symptoms of UA obstruction, alterations in physical exam, objective findings related to SDB and/or prematurity or family history of SDB. 2. Identification of comorbidities in CV system, CNS, nocturnal enuresis, growth delay or decreased QoL and conditions coexisting with SDB such as recurrent otitis media and history of tympanostomy tube placement, wheezing or asthma, metabolic syndrome or oral-motor dysfunction. 3. Recognition of factors predicting long-term persistence of SDB: obesity, male sex, obstructive AHI > 5/h, African-American ethnicity and persistent tonsillar hypertrophy and narrow mandible. 4. Objective diagnosis and assessment of SDB severity: PSG or RP is indicated in children at risk of SDB. (1) OSA definition 1: obstructive AHI $\geq$ 2/h or obstructive apnea index $\geq$ 1/h with SDB symptoms; (2) OSA definition 2: SDB symptoms and AHI $\geq$ 1/h. No alternative methods can substitute PSG but could be used in low resource settings: ambulatory PSG or RP, nocturnal oximetry, Pediatric Sleep Questionnaire or Sleep Clinical Record. 5. Indications for treatment of SDB: indicated when AHI > 5/h. When PSG or RP are not available, treatment is considered when positive oximetry or SDB questionnaires or morbidity is present. It is unclear whether should treat primary snoring (evaluation annually). 6. Stepwise treatment approach for SDB is usually implemented until complete resolution of SDB: (1) weight loss in overweight and obese children; (2) nasal corticosteroids and/or montelukast in non-obese and < 6 years children; (3) adenotonsillectomy in children with OSA and adenotonsillar hypertrophy; (4) rapid maxillary expansion or orthodontic appliances in children with OSA and maxillary constriction, retrognathia or malocclusion; (5) CPAP or NPPV when residual OSA after adenotonsillectomy or hypoventilation; (6) craniofacial surgery when syndromic craniofacial abnormalities; (7) tracheostomy in severe OSA when other nonsurgical or surgical interventions have failed or are contraindicated. 7. Recognition and management of persistent SDB: outcomes monitored after intervention are: symptoms, PSG (or RP, oximetry/capnography when not available), QoL, CV or CNS morbidity, enuresis and growth rate. PSG or RP should be performed, between 6–12 weeks after treatment, in children at risk of persistent OSA, after adenotonsillectomy, in children with persistent symptoms or children with mild OSA treated with corticosteroids and/or montelukast. PSG should be performed 12 months after rapid maxillary expansion and after 6 months when oral appliance treatment is selected. At least, one PSG or RP annually should be used to titrate CPAP or NPPV.

Table 3. *Cont.*

Guide	Diagnosis and Management of OSA in Children
American Academy of Pediatrics	This practice guideline focuses on uncomplicated childhood OSA, associated with adenotonsillar hypertrophy and/or obesity in an otherwise child who is being treated in the primary care setting. It comprises 8 key action statements. 1. Screening for OSA. If the child presents signs or symptoms of OSA, clinicians should perform medical history and physical examination. 2. Snoring and findings in the evaluation should lead to PSG (gold standard test) or alternative tests when PSG is not available (nocturnal video recording, nocturnal oximetry, daytime nap PSG or ambulatory PSG). 3. Adenotonsillectomy is recommended when the child is determined to have OSA and adenotonsillar hypertrophy (and do not have contraindication to surgery). If the child has OSA but not adenotonsillar hypertrophy other treatment should be considered. 4. Monitoring of high-risk patients undergoing adenotonsillectomy. 5. Reevaluation. Clinical reassessment should be performed in all patients with OSA for persisting symptoms after therapy to determine whether further treatment is required (6 to 8 weeks after treatment). 6. Clinicians should refer patients for CPAP management if symptoms persist after adenotonsillectomy or if it is not performed. 7. Weight loss is recommended in addition to other therapy if the child with OSA is overweight or obese. 8. Intranasal corticosteroids may be prescribed for children with mild OSA in whom surgery is contraindicated or have mild postoperative OSA (<5/h).

Abbreviations: OSA: obstructive sleep apnea; UA: upper airway; RP: respiratory polygraphy; PSG: polysomnography; AHI: apnea/hypopnea index; CPAP: continuous positive airway pressure; SDB: sleep-disordered breathing; CV: cardiovascular; CNS: central nervous system; QoL: quality of life; NPPV: non-invasive positive pressure ventilation.

Although medical history and physical examination are useful to screen and determine which patients are suspected of having OSA, the sensitivity and specificity are scarce. Thus, objective sleep tests are needed. The gold standard is overnight, attended, in-laboratory polysomnography (PSG), a complex test that records neurophysiological and cardiorespiratory variables. The American Academy of Sleep Medicine (AASM) in 2007 described the rules for the scoring of sleep and respiratory events in PSG recordings [4], last upgraded in February 2023, these being different for children than adults.

However, PSG may not be readily available, so alternative diagnostic tests can be performed: daytime nap PSG, ambulatory PSG, respiratory polygraphy (RP), nocturnal oximetry, the Pediatric Sleep Questionnaire, or nocturnal video recording. The complexity and limitations of PSG entail an increase in the development and validation of alternative methods for the diagnosis of OSA in children [42]. As an example, the European guideline accepts hospital RP as a valid alternative for the diagnosis of OSA in children and is considered an adequate screening technique when PSG is not available (Table 3).

1.6. Treatment

The goal of OSA treatment is complete resolution of SDB. This may require combining strategies (Figure 4), although the first-line treatment for OSA in children is adenotonsillar surgery [5] when adenotonsillar hypertrophy is present. Nonetheless, in the recent past, this treatment has been questioned. Recent publications have shown that the use of adenotonsillectomy in pediatric OSA patients may have variable results, reaching an AHI of 1 or less in about 50–70% of cases, but its efficacy decreases with risk factors such as age (<7 years), severe disease, chronic asthma or obesity. Persistent disease is present in 20–75% of children, with more than half having habitual snoring [7,43,44]. In addition, other surgical procedures may be performed in selected cases, such as septoplasty, uvulopharyngopalatoplasty, epiglottoplasty, glossopexy, and maxillomandibular surgery.

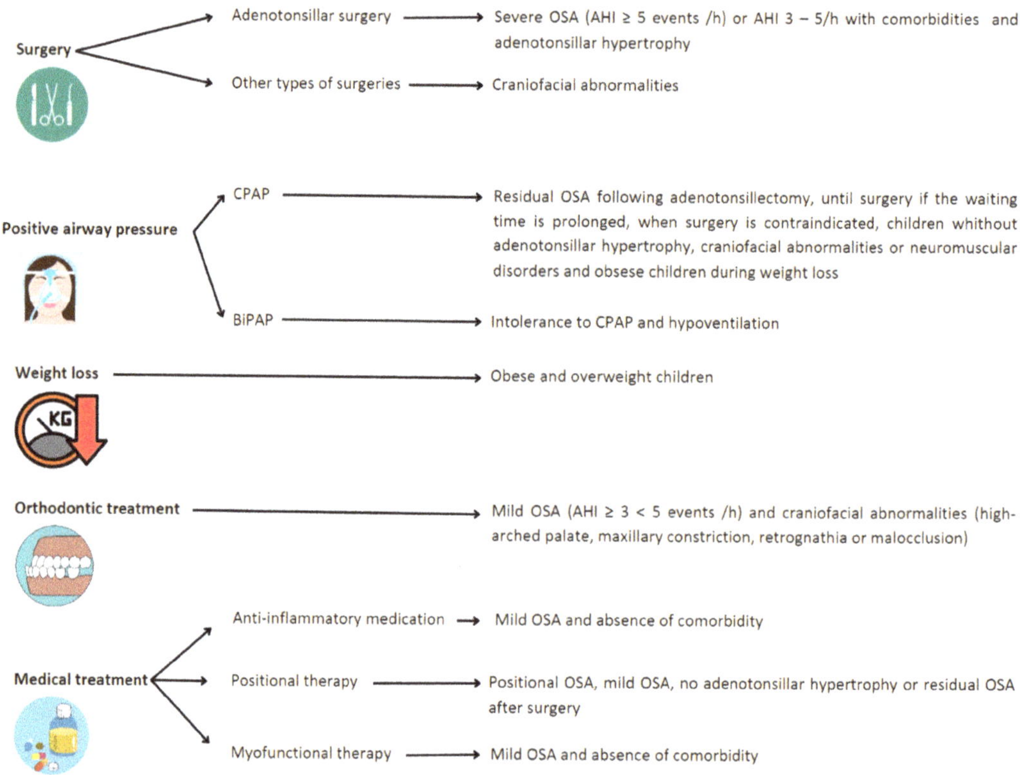

Figure 4. Recommended treatments depending on conditions leading to OSA in children. Abbreviations: OSA: obstructive sleep apnea; AHI: apnea/hypopnea index; CPAP: continuous positive airway pressure; BiPAP: bi-level positive airway pressure.

For those children with residual OSA following adenotonsillectomy or those in whom surgery is contraindicated or without adenotonsillar hypertrophy, positive airway pressure (PAP) therapies can be an effective treatment. The two types of PAP therapies prescribed in children to treat OSA are continuous positive airway pressure (CPAP) and bi-level positive airway pressure (Bi-PAP) [45]. CPAP is the most-commonly used PAP therapy, also recommended in children with craniofacial abnormalities or neuromuscular disorders [1,3,5]. The use of BiPAP is for patients intolerant to CPAP to treat nocturnal hypoventilation [46].

Positional therapy, as an alternative treatment, has been widely studied and relatively implemented in adults for the management of positional OSA. Positional OSA is defined when, spending more than 20% of sleep time in the supine position, the AHI in the supine position is at least double that in the non-supine position. This definition has not been adapted to the pediatric population and is, therefore, assumed in the child. In adults, positional therapy is incorporated in cases of mild–moderate OSA of positional origin and in those with severe OSA in order to lower CPAP pressure or when there is intolerance to first-line treatment. However, the indications in the pediatric population are not clearly established, and the scientific evidence is scarce. In this sense, it seems that children without tonsillar hypertrophy or with residual OSA could benefit from it, mainly in cases of obesity [47,48]. Therefore, randomized studies are necessary to establish the efficacy and indications of this type of therapy in the pediatric patient.

Weight management, orthodontic treatment, or medical therapy are offered as an alternative to surgery, especially in children with mild OSA [7,8,49] or when surgery is not indicated or contraindicated. There are data supporting that weight loss, if the child is overweight or obese, can improve OSA (hence, proposed to be considered first-line treatment in this population) [50]. Rapid maxillary expansion or orthodontic appliances are used to widen the palate and cause flattening of the palatal arch. On the other hand, medical therapies such as anti-inflammatory medications (nasal corticosteroid and/or oral montelukast) can also be used. There is little evidence about anti-inflammatory therapies in children. The results of randomized clinical trials evaluating the efficacy of intranasal corticosteroids for the treatment of OSA are not conclusive. Montelukast has short-term beneficial treatment effects for OSA in healthy, non-obese, surgically untreated children in terms of reducing the AHI, but the clinical relevance remains unclear [51,52]. Finally, myofunctional therapy has been accepted as a non-invasive treatment for OSA in children, as it may improve the AHI and oxygen saturation, at least after tonsillectomy or as an adjunct OSA treatment [53,54].

It is worth mentioning that obstructive SDB can be resolved spontaneously, particularly in children with mild OSA and adenotonsillar hypertrophy. Improvements may be due to the regression of lymphoid tissue or growth of the airway [55].

In summary, current data on the management of pediatric patients with OSA around the world, presented in this review, manifest important discrepancies and the need to be updated and homogenized. An agreed upon definition is needed with specific cutoff points to establish the diagnosis and levels of severity based on the associated risk and comorbidities. Clear diagnostic management algorithms must be settled upon in which it is defined when the simplified methods are useful in pediatric patients, in order to avoid underdiagnosis. It is necessary to identify prognostic markers that set up cutoff points for treatment indications based on objective impact. In addition, new metrics that better evaluate the disease could lead to the establishment of new protocols that improve the treatment management of the child. The objective is to explore new paradigms in the definition, diagnosis, and treatment of OSA in children. In this review, we analyzed the possible changes in the immediate future in this sense and what could be the new variables to establish the indication for the treatment of OSA.

2. Discussion

2.1. OSA Definition and Classification

While in adults, the definition of OSA is well established, this is not so clear in children, where these limits are between 1 and 3 events per hour, without an exact definition and important differences between guidelines. The new scientific evidence provided in recent years should be the basis for this new consensus and concrete definition. In the same way, the cut-off points for the levels of severity should be based on existing evidence about the associated risk related to the different levels of disease. In our opinion, an objective and concrete definition of OSA is needed, and the cutoff for severity levels should be stabilized based on the consequences of the illness.

2.2. OSA Diagnosis Algorithm

Related to the diagnostic management, PSG continues to be the gold standard test. Due to the scarcity of accredited sleep laboratories for children and important resources required to perform PSG, there has been a considerable effort to develop alternative diagnostic methods that are more widely available. However, more simplified tests have been developed, but insufficiently evaluated to be generally implemented in the pediatric population [56,57]. Hospital RP includes measures based on snoring, oronasal airflow, body position, chest movements, electrocardiography, and pulse oximeter saturation and has been used in Europe and Spain as a valid alternative to PSG [1,58]. Nonetheless, AASM's position is that these tests may underdiagnose OSA [59] and are less implemented in pediatric OSA. However, RP in adults has been demonstrated to be a very useful tool for underdiagnosis control when used in the right population and with an adequate interpretation. There have been many suggestions of performing home sleep apnea tests

(HSAT) [21] in children similarly to adults. However, the AASM position is that it is not recommended for the diagnosis of OSA in children [60].

Other modalities such as nocturnal oximetry recordings are a great possibility due to their low cost and their ease of use [61]. First, Brouillette and colleagues in 2000 [62] showed oximetric recordings as an approach for the identification of children with severe OSA. They concluded that a recording with at least three oxygen saturation (SpO_2) drops to less than 90% and three clusters of desaturation events could be considered diagnostic for OSA [62,63]. Warapongmanupong and Preutthipan in 2019 [64] suggested another oximetry parameter, where calculating the dispersion of SpO_2 could be useful in the initial investigation of OSA in children. A standard deviation of 1.06 or more could predict moderate–severe OSA with confidence.

Sleep questionnaires are helpful screening tools, but they cannot replace PSG [1]. To improve the diagnostic ability of clinical questionnaires, Villa et al. in 2013 [65] developed the Sleep Clinical Record (SCR), which included physical examination, patients' subjective symptoms, and the presence of inattention and hyperactivity. This score was designed to detect children with SDB and AHI > 1 episode/h. A recent study showed that snoring children with an SCR score above 8.25 could identify those with moderate-to-severe OSA [66]. The European guideline accepts this score as a valid method to diagnose OSA and determine its severity when PSG is not available [1].

In our opinion, an update to OSA diagnostic management in children should be performed, based on the actual epidemiological situation and technological advancement. The level of clinical suspicion should be established, based on medical history, and the utility of different diagnostic methods should be provided. These and other simplified diagnostic methods have been successfully applied to adults and should be evaluated and incorporated into the management of children to avoid misdiagnoses.

2.3. Specific OSA Biomarkers

Besides, diagnosis and severity classification are limited to the AHI in both children and adults, although this sole parameter does not reflect the heterogeneity of the disease. Additionally, severity groups are arbitrary and do not include consequences. In addition to the AHI, it should be recommended to take into account the immediate consequences (intermittent hypoxia (IH)), symptoms, obesity, and comorbidities present in the child.

A new stratification of OSA, based on a phenotype identification and with prognosis implications, is necessary. This path towards personalized medicine has had its advances and applications in adults, but it has not been replicated in the same way in children. It is known that groups of patients with similar characteristics can benefit from specific diagnostic and therapeutic management [67].

The limitations of the AHI are mainly related to the complexity of the disease. Besides, short- and long-term consequences make it necessary to develop new parameters that improve disease detection, predict associated health problems, and provide a better response to the impact of treatment. Individuals with similar AHI levels differ in the patterns of hypoxia, cardiac autonomic response, and respiratory arousal intensity that drive CV disease [68], reflecting that the AHI may not fully characterize the physiologic disturbances of OSA. For this reason, there is an increasing effort to develop new parameters that include these aspects.

First, measuring pro-inflammatory biomarkers in blood or urine such as interleukin 6, tumor necrosis factor alpha, or CRP may be relevant for a better characterization of the disease and may be helpful in evaluating the changes produced with different treatments [69].

Secondly, heart rate variability (HRV) has been recently described as an OSA-specific biomarker, with promising results in evaluating treatment efficacy in children [70–73]. HRV corresponds to variations in the heart rate (HR) or beat-beat time interval due to the modulation of the autonomic nervous system's activity. Because OSA sleep fragmentation and hypoxemia may increase the sympathetic nervous system's activity, HRV analysis could evaluate these characteristic variations produced by OSA [72,74].

Thirdly, high BP levels, even values close to normal, are a well-known modifiable factor for developing high blood pressure (HBP) [75] and increased CV risk [76,77] in the future. Thus, measuring BP in children could be considered a good CV risk marker. The pathophysiology of HBP in OSA depends on various factors, apart from increased sympathetic tone, such as peripheral vasoconstriction, increased renin–angiotensin–aldosterone activity, and proinflammatory responses [40,78], causing persistent increases in vascular resistance and altering BP [79]. Related to BP, it has been demonstrated that OSA in adults is a risk factor for developing a nondipping profile [80,81], produced when nocturnal BP decreases less than 10% of daytime BP. Moreover, this nondipping pattern is described as an independent risk factor for the development of adverse CV events [82]. However, much less is known in children, where only some studies have assessed the impact of OSA in circadian BP patterns [83–85]. Controversial results on this issue have questioned if nocturnal dipping is truly preserved in OSA children. Therefore, more studies in children are needed in order to evaluate the utility of this marker in the prevention of adverse consequences in adulthood.

Fourthly, IH occurs as a result of OSA and is considered the main deleterious factor involved in the consequences for CV risk associated with OSA. Conventional measures of hypoxia such as the oxygen desaturation index (ODI) or the percentage of time during sleep with oxygen saturation below 90% (T90) are commonly used to characterize OSA and its relationship with CV risk. These parameters have shown a better prediction of CV risk and mortality than the AHI [86,87]. Nevertheless, some of them are non-specific for OSA, reinforcing the idea of developing and validating more quantitative hypoxia metrics. In this sense, a novel OSA-specific biomarker, hypoxic burden (HB), has emerged for a better characterization of OSA, which focuses on the frequency, duration, and depth of the respiratory events. HB is defined as the total area under the desaturation curve related to the respiratory event. The value is obtained by adding the individual desaturation areas and dividing the total area by the duration of sleep, in units of %min/h [88]. This parameter has been shown to be significantly associated with adverse CV outcomes in recent studies in adults [89]; however, the search within this issue has not yet been evaluated in children. Measuring HB in children with OSA may be important and of interest, as it could better predict the risk attributed to the disease and could improve the choice of treatment. Besides, HB measurement requires only recording of airflow and oxygen saturation signals, so it could be easily obtained by simplified methods, facilitating its implementation in clinical practice.

Therefore, an effort to phenotype and establish the value of the different biomarkers on the way to personalized medicine is needed in the OSA pediatric population.

2.4. Treatment

In reference to treatment, the lack of adenotonsillectomy effectiveness in some groups of patients reflects that they may have a different response to treatment and/or a different prognosis. Furthermore, OSA treatments are not exclusive, and all the therapeutic options should be considered to approach treatment individually. For this, we highlight the importance of looking for the reversible cause of the disease, so that future risk can be prevented. The most-common cause is adenotonsillar hypertrophy, but other risk factors such as obesity and craniofacial dysmorphism should be taken into account for the correct management of the patient. In this sense, the treatment of obesity in adult patients is a cornerstone in its control, becoming ahead of CPAP treatment in overweight/obese patients in the latest consensus document [90]. It is true that the presence of obesity in children is less prevalent, but there has been a significant increase in recent times. This fact contributes to the higher frequency of OSA cases in the absence of tonsillar hypertrophy and residual OSA after tonsillectomy due to overweight/obesity. However, the impact of obesity treatment in children has not been studied to the same extent as in adults, nor has its role in the diagnostic algorithm been established. Other different possibilities that could be used include the use of CPAP or orthodontic treatment in patients with certain craniofacial abnormalities. The early use of these techniques could prevent the risk of suffering from OSA.

On the other hand, the development of innovative OSA parameters related to CV risk in the pediatric population could initiate new paradigms in the management of children with OSA, as they could better predict and determine which patients would benefit from treatments. One possibility could be measuring the HB in the pediatric population as a substitute of the AHI in the diagnosis and classification of the severity groups of OSA. Relating this parameter to BP values could indicate which groups would benefit from OSA treatment in terms of the prevention of future CV risk.

Additionally, the reevaluation of children should be performed after treatment to determine whether further treatment is necessary, especially in children at risk of residual OSA. It is not clear when the ideal moment for this reassessment is, specific recommendations based on different available therapies being necessary.

In summary, OSA treatment in children should focus on considering these two important aspects: reversible causes of the disease and measurements defining which patients would benefit from them. New clinical trials would be needed to assess the feasibility of the implementation in clinical practice of these innovative insights of the management of OSA in children.

3. Conclusions

OSA in the pediatric population has important limitations that should be updated.

First, a specific and consensus definition should be established. Second, the role of simplified methods in the diagnosis algorithm should be considered. Third, new biomarkers for risk identification are needed in this population. Finally, personalized medicine should be implemented in this population.

Author Contributions: Conceptualization, E.S.-P., C.C. and O.M.; methodology, E.S.-P., C.C. and O.M.; validation, E.S.-P., C.C. and O.M.; investigation, E.S.-P., C.C., M.C.-G., S.R.-P., S.L.-M., E.L., I.C.-P., M.S.-d.-l.-T., F.G.-R. and O.M.; resources, O.M.; writing—original draft preparation, E.S.-P. and C.C.; writing—review and editing, O.M.; supervision, O.M.; funding acquisition, O.M. All authors have read and agreed to the published version of the manuscript.

Funding: This work was supported by funds received by the ISCIII (PI18/00565 and PI22/01653), the Spanish Respiratory Society (Sociedad Española de Neumología y Cirugía Torácica—SEPAR) (535-2018 and 1073-2020), and an unconditioned research grant from Menarini Laboratories and NEUMOMADRID (Sociedad Madrileña de Neumología y Cirugía Torácica).

Institutional Review Board Statement: Not applicable.

Informed Consent Statement: Not applicable.

Data Availability Statement: Not applicable.

Conflicts of Interest: The authors declare no conflict of interest.

Abbreviations

Sleep-disordered breathing (SDB); upper airway (UA); obstructive sleep apnea (OSA); apnea/hypopnea index (AHI); cardiovascular (CV); C-reactive protein (CRP); blood pressure (BP); polysomnography (PSG); American Academy of Sleep Medicine (AASM); respiratory polygraphy (RP); positive airway pressure (PAP); continuous positive airway pressure (CPAP); bi-level positive airway pressure (Bi-PAP); home sleep apnea test (HSAT); oxygen saturation (SpO_2); Sleep Clinical Record (SCR); intermittent hypoxia (IH); heart rate variability (HRV); heart rate (HR); high blood pressure (HBP); oxygen desaturation index (ODI); oxygen saturation below 90% (T90); hypoxic burden (HB).

References

1. Kaditis, A.G.; Alonso Alvarez, M.L.; Boudewyns, A.; Alexopoulos, E.I.; Ersu, R.; Joosten, K.; Larramona, H.; Miano, S.; Narang, I.; Trang, H.; et al. Obstructive Sleep Disordered Breathing in 2- to 18-Year-Old Children: Diagnosis and Management. *Eur. Respir. J.* **2016**, *47*, 69–94. [CrossRef] [PubMed]
2. Gozal, D.; O'Brien, L.M. Snoring and Obstructive Sleep Apnoea in Children: Why Should We Treat? *Paediatr. Respir. Rev.* **2004**, *5*, S371–S376. [CrossRef] [PubMed]
3. Marcus, C.L.; Brooks, L.J.; Ward, S.D.; Draper, K.A.; Gozal, D.; Halbower, A.C.; Jones, J.; Lehmann, C.; Schechter, M.S.; Sheldon, S.; et al. Diagnosis and Management of Childhood Obstructive Sleep Apnea Syndrome. *Pediatrics* **2012**, *130*, e714–e755. [CrossRef]
4. Iber, C.; Ancoli-Israel, S.; Chesson, A.L.; Quan, S.F. *The AASM Manual for the Scoring of Sleep and Associated Events: Rules, Terminology and Technical Specifications*; American Academy of Sleep Medicine: Westchester, IL, USA, 2007; Volume 1.
5. Luz Alonso-Álvarez, M.; Canet, T.; Cubell-Alarco, M.; Estivill, E.; Fernández-Julián, E.; Gozal, D.; Jurado-Luque, M.J.; Lluch-Roselló, M.A.; Martínez-Pérez, F.; Merino-Andreu, M.; et al. Documento de consenso del síndrome de apneas-hipopneas durante el sueño en niños (versión completa). *Arch. Bronconeumol.* **2011**, *47*, 2–18. [CrossRef]
6. Witmans, M.; Tablizo, M.A. Current Concepts in Pediatric Obstructive Sleep Apnea. *Children* **2023**, *10*, 480. [CrossRef]
7. Bitners, A.C.; Arens, R. Evaluation and Management of Children with Obstructive Sleep Apnea Syndrome. *Lung* **2020**, *198*, 257–270. [CrossRef]
8. Dehlink, E.; Tan, H.-L. Update on Paediatric Obstructive Sleep Apnoea. *J. Thorac. Dis.* **2016**, *8*, 224–235.
9. Lo Bue, A.; Salvaggio, A.; Insalaco, G. Obstructive Sleep Apnea in Developmental Age. A Narrative Review. *Eur. J. Pediatr.* **2020**, *179*, 357–365. [CrossRef]
10. Gulotta, G.; Iannella, G.; Vicini, C.; Polimeni, A.; Greco, A.; de Vincentiis, M.; Visconti, I.C.; Meccariello, G.; Cammaroto, G.; De Vito, A.; et al. Risk Factors for Obstructive Sleep Apnea Syndrome in Children: State of the Art. *Int. J. Environ. Res. Public Health* **2019**, *16*, 3235. [CrossRef]
11. Al-Shamrani, A.; Alharbi, A.S. Diagnosis and Management of Childhood Sleep-Disordered Breathing: Clinical Approach. *Saudi Med. J.* **2020**, *41*, 916–929. [CrossRef]
12. Lin, S.; Su, Y.; Wu, Y.; Chang, J.Z.; Tu, Y. Management of Paediatric Obstructive Sleep Apnoea: A Systematic Review and Network Meta-analysis. *Int. J. Paediatr. Dent.* **2020**, *30*, 156–170. [CrossRef] [PubMed]
13. Lam, Y.; Chan, E.Y.T.; Ng, D.K.; Chan, C.; Cheung, J.M.Y.; Leung, S.; Chow, P.; Kwok, K. The Correlation Among Obesity, Apnea-Hypopnea Index, and Tonsil Size in Children. *Chest* **2006**, *130*, 1751–1756. [CrossRef] [PubMed]
14. Wang, J.; Zhao, Y.; Yang, W.; Shen, T.; Xue, P.; Yan, X.; Chen, D.; Qiao, Y.; Chen, M.; Ren, R.; et al. Correlations between Obstructive Sleep Apnea and Adenotonsillar Hypertrophy in Children of Different Weight Status. *Sci. Rep.* **2019**, *9*, 11455. [CrossRef] [PubMed]
15. Li, J.; Yang, Q.; Xu, Y.; Han, F.; Zhao, J. Research progress on correlation between childhood obesity and obstructive sleep apnea. *Lin Chuang Er Bi Yan Hou Tou Jing Wai Ke Za Zhi* **2023**, *37*, 318–322. [CrossRef]
16. Narang, I.; Mathew, J.L. Childhood Obesity and Obstructive Sleep Apnea. *J. Nutr. Metab.* **2012**, *2012*, 134202. [CrossRef]
17. Lee, J.H.; Cho, J. Sleep and Obesity. *Sleep Med. Clin.* **2022**, *17*, 111–116. [CrossRef]
18. Goodwin, J.L.; Vasquez, M.M.; Silva, G.E.; Quan, S.F. Incidence and Remission of Sleep-Disordered Breathing and Related Symptoms in 6- to 17-Year Old Children—The Tucson Children's Assessment of Sleep Apnea Study. *J. Pediatr.* **2010**, *157*, 57–61. [CrossRef]
19. Savini, S.; Ciorba, A.; Bianchini, C.; Stomeo, F.; Corazzi, V.; Vicini, C.; Pelucchi, S. Assessment of Obstructive Sleep Apnoea (OSA) in Children: An Update. *Acta Otorhinolaryngol. Ital.* **2019**, *39*, 289–297. [CrossRef]
20. Zaffanello, M.; Piacentini, G.; Lippi, G.; Fanos, V.; Gasperi, E.; Nosetti, L. Obstructive Sleep-Disordered Breathing, Enuresis and Combined Disorders in Children: Chance or Related Association? *Swiss Med. Wkly.* **2017**, *147*, w14400. [CrossRef]
21. Penzel, T.; Hornero, R. (Eds.) *Advances in the Diagnosis and Treatment of Sleep Apnea: Filling the Gap between Physicians and Engineers*; Advances in Experimental Medicine and Biology; Springer International Publishing: Cham, Switzerland, 2022; Volume 1384, ISBN 978-3-031-06412-8.
22. Beebe, D.W.; Rausch, J.; Byars, K.C.; Lanphear, B.; Yolton, K. Persistent Snoring in Preschool Children: Predictors and Behavioral and Developmental Correlates. *Pediatrics* **2012**, *130*, 382–389. [CrossRef]
23. Bucks, R.S.; Olaithe, M.; Eastwood, P. Neurocognitive Function in Obstructive Sleep Apnoea: A Meta-Review: Cognitive Function in OSA: A Meta-Review. *Respirology* **2013**, *18*, 61–70. [CrossRef] [PubMed]
24. Urschitz, M.S.; Guenther, A.; Eggebrecht, E.; Wolff, J.; Urschitz-Duprat, P.M.; Schlaud, M.; Poets, C.F. Snoring, Intermittent Hypoxia and Academic Performance in Primary School Children. *Am. J. Respir. Crit. Care Med.* **2003**, *168*, 464–468. [CrossRef] [PubMed]
25. Galland, B.; Spruyt, K.; Dawes, P.; McDowall, P.S.; Elder, D.; Schaughency, E. Sleep Disordered Breathing and Academic Performance: A Meta-Analysis. *Pediatrics* **2015**, *136*, e934–e946. [CrossRef] [PubMed]
26. Brockmann, P.E.; Gozal, D. Neurocognitive Consequences in Children with Sleep Disordered Breathing: Who Is at Risk? *Children* **2022**, *9*, 1278. [CrossRef] [PubMed]
27. Bonuck, K.A.; Freeman, K.; Henderson, J. Growth and Growth Biomarker Changes after Adenotonsillectomy: Systematic Review and Meta-Analysis. *Arch. Dis. Child.* **2008**, *94*, 83–91. [CrossRef] [PubMed]

28. Esteller, E.; Villatoro, J.C.; Agüero, A.; Lopez, R.; Matiñó, E.; Argemi, J.; Girabent-Farrés, M. Obstructive Sleep Apnea Syndrome and Growth Failure. *Int. J. Pediatr. Otorhinolaryngol.* **2018**, *108*, 214–218. [CrossRef]
29. Gümüssoy, M.; Atmaca, S.; Bilgici, B.; Ünal, R. Changes in IGF-I, IGFBP-3 and Ghrelin Levels after Adenotonsillectomy in Children with Sleep Disordered Breathing. *Int. J. Pediatr. Otorhinolaryngol.* **2009**, *73*, 1653–1656. [CrossRef]
30. Bhattacharjee, R.; Kheirandish-Gozal, L.; Pillar, G.; Gozal, D. Cardiovascular Complications of Obstructive Sleep Apnea Syndrome: Evidence from Children. *Prog. Cardiovasc. Dis.* **2009**, *51*, 416–433. [CrossRef]
31. Kang, K.-T.; Weng, W.-C.; Lee, P.-L.; Hsu, W.-C. C-Reactive Protein in Children with Obstructive Sleep Apnea and Effects of Adenotonsillectomy. *Auris Nasus Larynx* **2022**, *49*, 92–99. [CrossRef]
32. Smith, D.F.; Amin, R.S. OSA and Cardiovascular Risk in Pediatrics. *Chest* **2019**, *156*, 402–413. [CrossRef]
33. Amin, R.S.; Kimball, T.R.; Kalra, M.; Jeffries, J.L.; Carroll, J.L.; Bean, J.A.; Witt, S.A.; Glascock, B.J.; Daniels, S.R. Left Ventricular Function in Children with Sleep-Disordered Breathing. *Am. J. Cardiol.* **2005**, *95*, 801–804. [CrossRef]
34. Nemati, S.; Aghajankhah, M.; Banan, R.; Haddadi, S.; Mehri, M.; Aghsaghloo, V.; Leili, E.K. The Effects of Adeno/Tonsillectomy on Cardiopulmonary Function Based on Echocardiography Indices in Children with Primary Snoring and Mild Obstructive Sleep Apnea. *Am. J. Otolaryngol.* **2022**, *43*, 103317. [CrossRef] [PubMed]
35. Sameema, V.V.; Soni, K.; Deora, S.; Sharma, J.B.; Choudhury, B.; Kaushal, D.; Chhabra, S.; Goyal, A. Assessment of Preoperative and Postoperative Cardiac Function in Children with Adenotonsillar Hypertrophy: A Prospective Cohort Study. *Eur. Arch. Otorhinolaryngol.* **2022**, *279*, 3013–3019. [CrossRef] [PubMed]
36. Giuca, M.R.; Carli, E.; Lardani, L.; Pasini, M.; Miceli, M.; Fambrini, E. Pediatric Obstructive Sleep Apnea Syndrome: Emerging Evidence and Treatment Approach. *Sci. World J.* **2021**, *2021*, 5591251. [CrossRef]
37. Redline, S.; Storfer-Isser, A.; Rosen, C.L.; Johnson, N.L.; Kirchner, H.L.; Emancipator, J.; Kibler, A.M. Association between Metabolic Syndrome and Sleep-Disordered Breathing in Adolescents. *Am. J. Respir. Crit. Care Med.* **2007**, *176*, 401–408. [CrossRef] [PubMed]
38. Menzies, B.; Teng, A.; Burns, M.; Lah, S. Neurocognitive Outcomes of Children with Sleep Disordered Breathing: A Systematic Review with Meta-Analysis. *Sleep Med. Rev.* **2022**, *63*, 101629. [CrossRef]
39. Lagravère, M.O.; Zecca, P.A.; Caprioglio, A.; Fastuca, R. Metabolic Effects of Treatment in Patients with Obstructive Sleep Apnea: A Systematic Review. *Minerva Pediatr.* **2019**, *71*, 380–389. [CrossRef]
40. Ai, S.; Li, Z.; Wang, S.; Chen, S.; Chan, J.W.; Au, C.T.; Bao, Y.; Li, A.M.; Zhang, J.; Chan, K.C.-C.; et al. Blood Pressure and Childhood Obstructive Sleep Apnea: A Systematic Review and Meta-Analysis. *Sleep Med. Rev.* **2022**, *65*, 101663. [CrossRef]
41. Thomas, S.; Patel, S.; Gummalla, P.; Tablizo, M.A.; Kier, C. You Cannot Hit Snooze on OSA: Sequelae of Pediatric Obstructive Sleep Apnea. *Children* **2022**, *9*, 261. [CrossRef]
42. Kang, M.; Mo, F.; Witmans, M.; Santiago, V.; Tablizo, M.A. Trends in Diagnosing Obstructive Sleep Apnea in Pediatrics. *Children* **2022**, *9*, 306. [CrossRef]
43. Huang, Y.-S.; Guilleminault, C. Pediatric Obstructive Sleep Apnea: Where Do We Stand? In *Advances in Oto-Rhino-Laryngology*; Lin, H.-C., Ed.; S. Karger AG: Basel, Switzerland, 2017; Volume 80, pp. 136–144. ISBN 978-3-318-06064-5.
44. Pizarro, G.U.; Costa, E.L.d.B.; Pradella-Hallinan, M.; Meurer, A.T.d.O.; Moreira, G.A.; Fujita, R.R. Efficacy of Adenotonsillectomy in the Treatment of Obstructive Apnea in Children: A 2-Year Follow-Up. *Int. J. Pediatr. Otorhinolaryngol.* **2023**, *166*, 111462. [CrossRef] [PubMed]
45. Parmar, A.; Baker, A.; Narang, I. Positive Airway Pressure in Pediatric Obstructive Sleep Apnea. *Paediatr. Respir. Rev.* **2019**, *31*, 43–51. [CrossRef] [PubMed]
46. Machaalani, R.; Evans, C.A.; Waters, K.A. Objective Adherence to Positive Airway Pressure Therapy in an Australian Paediatric Cohort. *Sleep Breath.* **2016**, *20*, 1327–1336. [CrossRef] [PubMed]
47. Xiao, L.; Baker, A.; Voutsas, G.; Massicotte, C.; Wolter, N.E.; Propst, E.J.; Narang, I. Positional Device Therapy for the Treatment of Positional Obstructive Sleep Apnea in Children: A Pilot Study. *Sleep Med.* **2021**, *85*, 313–316. [CrossRef] [PubMed]
48. Tholen, K.; Meier, M.; Kloor, J.; Friedman, N. Persistent OSA in Obese Children: Does Body Position Matter? *J. Clin. Sleep Med.* **2021**, *17*, 227–232. [CrossRef]
49. Brockbank, J.C. Update on Pathophysiology and Treatment of Childhood Obstructive Sleep Apnea Syndrome. *Paediatr. Respir. Rev.* **2017**, *24*, 21–23. [CrossRef]
50. Andersen, I.G.; Holm, J.-C.; Homøe, P. Impact of Weight-Loss Management on Children and Adolescents with Obesity and Obstructive Sleep Apnea. *Int. J. Pediatr. Otorhinolaryngol.* **2019**, *123*, 57–62. [CrossRef]
51. Nixon, G.M.; Perrett, K.P. Limited Evidence for Anti-inflammatory Medications for Obstructive Sleep Apnoea in Children. *J. Paediatr. Child Health* **2021**, *57*, 2019–2021. [CrossRef]
52. Kuhle, S.; Urschitz, M.S. Anti-Inflammatory Medications for the Treatment of Pediatric Obstructive Sleep Apnea. *Paediatr. Respir. Rev.* **2020**, *34*, 35–36. [CrossRef]
53. Zhang, F.; Tian, Z.; Shu, Y.; Zou, B.; Yao, H.; Li, S.; Li, Q. Efficiency of Oro-Facial Myofunctional Therapy in Treating Obstructive Sleep Apnoea: A Meta-Analysis of Observational Studies. *J. Oral Rehabil.* **2022**, *49*, 734–745. [CrossRef]
54. Bandyopadhyay, A.; Kaneshiro, K.; Camacho, M. Effect of Myofunctional Therapy on Children with Obstructive Sleep Apnea: A Meta-Analysis. *Sleep Med.* **2020**, *75*, 210–217. [CrossRef]
55. Marcus, C.L.; Moore, R.H.; Rosen, C.L.; Giordani, B.; Garetz, S.L.; Taylor, H.G.; Mitchell, R.B.; Amin, R.; Katz, E.S.; Arens, R.; et al. A Randomized Trial of Adenotonsillectomy for Childhood Sleep Apnea. *N. Engl. J. Med.* **2013**, *368*, 2366–2376. [CrossRef]

56. Alonso-Álvarez, M.L.; Terán-Santos, J.; Ordax Carbajo, E.; Cordero-Guevara, J.A.; Navazo-Egüia, A.I.; Kheirandish-Gozal, L.; Gozal, D. Reliability of Home Respiratory Polygraphy for the Diagnosis of Sleep Apnea in Children. *Chest* **2015**, *147*, 1020–1028. [CrossRef] [PubMed]
57. Oceja, E.; Rodríguez, P.; Jurado, M.; Luz Alonso, M.; del Río, G.; Villar, M.; Mediano, O.; Martínez, M.; Juarros, S.; Merino, M.; et al. Validity and Cost-Effectiveness of Pediatric Home Respiratory Polygraphy for the Diagnosis of Obstructive Sleep Apnea in Children: Rationale, Study Design, and Methodology. *Methods Protoc.* **2021**, *4*, 9. [CrossRef]
58. Alonso Alvarez, M.L.; Terán Santos, J.; Cordero Guevara, J.A.; Navazo Egüia, A.I.; Ordax Carbajo, E.; Masa Jiménez, J.F.; Pelayo, R. Reliability of respiratory polygraphy for the diagnosis of sleep apnea-hypopnea syndrome in children. *Arch. Bronconeumol.* **2008**, *44*, 318–323. [CrossRef] [PubMed]
59. Tan, H.-L.; Gozal, D.; Ramirez, H.M.; Bandla, H.P.R.; Kheirandish-Gozal, L. Overnight Polysomnography versus Respiratory Polygraphy in the Diagnosis of Pediatric Obstructive Sleep Apnea. *Sleep* **2014**, *37*, 255–260. [CrossRef]
60. Kirk, V.; Baughn, J.; D'Andrea, L.; Friedman, N.; Galion, A.; Garetz, S.; Hassan, F.; Wrede, J.; Harrod, C.G.; Malhotra, R.K. American Academy of Sleep Medicine Position Paper for the Use of a Home Sleep Apnea Test for the Diagnosis of OSA in Children. *J. Clin. Sleep Med.* **2017**, *13*, 1199–1203. [CrossRef] [PubMed]
61. Kaditis, A.; Kheirandish-Gozal, L.; Gozal, D. Pediatric OSAS: Oximetry Can Provide Answers When Polysomnography Is Not Available. *Sleep Med. Rev.* **2016**, *27*, 96–105. [CrossRef]
62. Brouillette, R.T.; Morielli, A.; Leimanis, A.; Waters, K.A.; Luciano, R.; Ducharme, F.M. Nocturnal Pulse Oximetry as an Abbreviated Testing Modality for Pediatric Obstructive Sleep Apnea. *Pediatrics* **2000**, *105*, 405–412. [CrossRef] [PubMed]
63. Gozal, D.; Kheirandish-Gozal, L.; Kaditis, A.G. Home Sleep Testing for the Diagnosis of Pediatric Obstructive Sleep Apnea: The Times They Are a Changing . . . ! *Curr. Opin. Pulm. Med.* **2015**, *21*, 563–568. [CrossRef]
64. Warapongmanupong, S.; Preutthipan, A. Can Standard Deviation of Overnight Pulse Oximetry Be Used to Screen Childhood Obstructive Sleep Apnea. *Int. J. Pediatr. Otorhinolaryngol.* **2019**, *119*, 27–31. [CrossRef]
65. Villa, M.P.; Paolino, M.C.; Castaldo, R.; Vanacore, N.; Rizzoli, A.; Miano, S.; Del Pozzo, M.; Montesano, M. Sleep Clinical Record: An Aid to Rapid and Accurate Diagnosis of Paediatric Sleep Disordered Breathing. *Eur. Respir. J.* **2013**, *41*, 1355–1361. [CrossRef] [PubMed]
66. Mylona, A.M.; Rapti, G.; Vavougios, G.; Lachanas, V.A.; Liakos, P.; Skoulakis, C.; Kaditis, A.G.; Gourgoulianis, K.; Alexopoulos, E.I. Accuracy of the Sleep Clinical Record for the Diagnosis of Pediatric Moderate-to-Severe Obstructive Sleep Apnea Syndrome. *Sleep Breath.* **2022**, *26*, 763–769. [CrossRef] [PubMed]
67. Romero-Peralta, S.; García-Rio, F.; Resano Barrio, P.; Viejo-Ayuso, E.; Izquierdo, J.L.; Sabroso, R.; Castelao, J.; Fernández Francés, J.; Mediano, O. Defining the Heterogeneity of Sleep Apnea Syndrome: A Cluster Analysis with Implications for Patient Management. *Arch. Bronconeumol.* **2022**, *58*, 125–134. [CrossRef]
68. Redline, S.; Azarbarzin, A.; Peker, Y. Obstructive Sleep Apnoea Heterogeneity and Cardiovascular Disease. *Nat. Rev. Cardiol.* **2023**. [CrossRef]
69. Tiboc-Schnell, C.N.; Filip, G.A.; Bolboaca, S.D.; Decea, N.; Chereches Panta, P.; Iacob, D.; Mihut, G.; Marin, A.; Man, S.C. Biomarkers of Pediatric Obstructive Sleep Apnea Syndrome and the Assessment of Quality of Life before and after Adenotonsillectomy. *J. Physiol. Pharmacol.* **2021**, *72*, 583–594. [CrossRef]
70. Martín-Montero, A.; Gutiérrez-Tobal, G.C.; Kheirandish-Gozal, L.; Vaquerizo-Villar, F.; Álvarez, D.; del Campo, F.; Gozal, D.; Hornero, R. Heart Rate Variability as a Potential Biomarker of Pediatric Obstructive Sleep Apnea Resolution. *Sleep* **2022**, *45*, zsab214. [CrossRef]
71. Rodríguez-Núñez, I.; Rodríguez-Romero, N.; Álvarez, A.; Zambrano, L.; Luciano da Veiga, G.; Romero, F.; Rodríguez-Núñez, I.; Rodríguez-Romero, N.; Álvarez, A.; Zambrano, L.; et al. Variabilidad del ritmo cardíaco en pediatría: Aspectos metodológicos y aplicaciones clínicas. *Arch. Cardiol. Mex.* **2022**, *92*, 6659. [CrossRef]
72. Lee, L.-A.; Chuang, H.-H.; Hsieh, H.-S.; Wang, C.-Y.; Chuang, L.-P.; Li, H.-Y.; Fang, T.-J.; Huang, Y.-S.; Lee, G.-S.; Yang, A.C.; et al. Using Sleep Heart Rate Variability to Investigate the Sleep Quality in Children with Obstructive Sleep Apnea. *Front. Public Health* **2023**, *11*, 1103085. [CrossRef]
73. Martín-Montero, A.; Armañac-Julián, P.; Gil, E.; Kheirandish-Gozal, L.; Álvarez, D.; Lázaro, J.; Bailón, R.; Gozal, D.; Laguna, P.; Hornero, R.; et al. Pediatric Sleep Apnea: Characterization of Apneic Events and Sleep Stages Using Heart Rate Variability. *Comput. Biol. Med.* **2023**, *154*, 106549. [CrossRef] [PubMed]
74. O'Driscoll, D.M.; Horne, R.S.C.; Davey, M.J.; Hope, S.A.; Anderson, V.; Trinder, J.; Walker, A.M.; Nixon, G.M. Increased Sympathetic Activity in Children with Obstructive Sleep Apnea: Cardiovascular Implications. *Sleep Med.* **2011**, *12*, 483–488. [CrossRef]
75. Tirosh, A.; Afek, A.; Rudich, A.; Percik, R.; Gordon, B.; Ayalon, N.; Derazne, E.; Tzur, D.; Gershnabel, D.; Grossman, E.; et al. Progression of Normotensive Adolescents to Hypertensive Adults: A Study of 26 980 Teenagers. *Hypertension* **2010**, *56*, 203–209. [CrossRef] [PubMed]
76. Tzeng, N.-S.; Chung, C.-H.; Chang, H.-A.; Chang, C.-C.; Lu, R.-B.; Yeh, H.-W.; Chiang, W.-S.; Kao, Y.-C.; Chang, S.-Y.; Chien, W.-C. Obstructive Sleep Apnea in Children and Adolescents and the Risk of Major Adverse Cardiovascular Events: A Nationwide Cohort Study in Taiwan. *J. Clin. Sleep Med.* **2019**, *15*, 275–283. [CrossRef] [PubMed]
77. Bassareo, P.P.; Calcaterra, G.; Sabatino, J.; Oreto, L.; Ciliberti, P.; Perrone, M.; Martino, F.; D'Alto, M.; Chessa, M.; DI Salvo, G.; et al. Primary and Secondary Paediatric Hypertension. *J. Cardiovasc. Med.* **2023**, *24*, e77–e85. [CrossRef]

78. Ahmad, M.; Makati, D.; Akbar, S. Review of and Updates on Hypertension in Obstructive Sleep Apnea. *Int. J. Hypertens.* **2017**, *2017*, 1848375. [CrossRef] [PubMed]
79. Shanmugam, A.; Binney, Z.O.; Voyles, C.B.; Bouldin, E.; Raol, N.P. Impact of OSA Treatment Success on Changes in Hypertension and Obesity: A Retrospective Cohort Study. *Sleep Med.* **2023**, *101*, 205–212. [CrossRef]
80. Pinilla, L.; Benítez, I.D.; Gracia-Lavedan, E.; Torres, G.; Minguez, O.; Aguilà, M.; Targa, A.; Dalmases, M.; Mediano, O.; Masa, J.F.; et al. Polysomnographic Characterization of Circadian Blood Pressure Patterns in Patients with Obstructive Sleep Apnea. *Sleep* **2023**, *46*, zsad031. [CrossRef]
81. Crinion, S.J.; Kleinerova, J.; Kent, B.; Nolan, G.; Taylor, C.T.; Ryan, S.; McNicholas, W.T. Non-Dipping Nocturnal Blood Pressure Correlates with Obstructive Sleep Apnoea Severity in Normotensive Subjects and May Reverse with Therapy. *ERJ Open Res.* **2021**, *7*, 00338-2021. [CrossRef]
82. Zweiker, R.; Eber, B.; Schumacher, M.; Toplak, H.; Klein, W. "Non-Dipping" Related to Cardiovascular Events in Essential Hypertensive Patients. *Acta Med. Austriaca* **1994**, *21*, 86–89. [PubMed]
83. Horne, R.S.C.; Yang, J.S.C.; Walter, L.M.; Richardson, H.L.; O'Driscoll, D.M.; Foster, A.M.; Wong, S.; Ng, M.L.; Bashir, F.; Patterson, R.; et al. Nocturnal Dipping Is Preserved in Children with Sleep Disordered Breathing Regardless of Its Severity: Nocturnal Dipping and Sleep Disordered Breathing. *Pediatr. Pulmonol.* **2013**, *48*, 1127–1134. [CrossRef]
84. Nisbet, L.C.; Nixon, G.M.; Yiallourou, S.R.; Biggs, S.N.; Davey, M.J.; Trinder, J.; Walter, L.M.; Horne, R.S.C. Sleep-Disordered Breathing Does Not Affect Nocturnal Dipping, as Assessed by Pulse Transit Time, in Preschool Children: Evidence for Early Intervention to Prevent Adverse Cardiovascular Effects? *Sleep Med.* **2014**, *15*, 464–471. [CrossRef] [PubMed]
85. Chan, K.C.; Au, C.T.; Hui, L.L.; Wing, Y.K.; Li, A.M. Childhood OSA Is an Independent Determinant of Blood Pressure in Adulthood: Longitudinal Follow-up Study. *Thorax* **2020**, *75*, 422–431. [CrossRef] [PubMed]
86. Xu, P.-H.; Fong, D.Y.T.; Lui, M.M.S.; Lam, D.C.L.; Ip, M.S.M. Cardiovascular Outcomes in Obstructive Sleep Apnoea and Implications of Clinical Phenotyping on Effect of CPAP Treatment. *Thorax* **2023**, *78*, 76–84. [CrossRef] [PubMed]
87. Cao, W.; Luo, J.; Xiao, Y. A Review of Current Tools Used for Evaluating the Severity of Obstructive Sleep Apnea. *Nat. Sci. Sleep* **2020**, *12*, 1023–1031. [CrossRef]
88. Azarbarzin, A.; Sands, S.A.; Stone, K.L.; Taranto-Montemurro, L.; Messineo, L.; Terrill, P.I.; Ancoli-Israel, S.; Ensrud, K.; Purcell, S.; White, D.P.; et al. The Hypoxic Burden of Sleep Apnoea Predicts Cardiovascular Disease-Related Mortality: The Osteoporotic Fractures in Men Study and the Sleep Heart Health Study. *Eur. Heart J.* **2019**, *40*, 1149–1157. [CrossRef]
89. Martinez-Garcia, M.A.; Sánchez-de-la-Torre, M.; White, D.P.; Azarbarzin, A. Hypoxic Burden in Obstructive Sleep Apnea: Present and Future. *Arch. Bronconeumol.* **2023**, *59*, 36–43. [CrossRef]
90. Mediano, O.; González Mangado, N.; Montserrat, J.M.; Alonso-Álvarez, M.L.; Almendros, I.; Alonso-Fernández, A.; Barbé, F.; Borsini, E.; Caballero-Eraso, C.; Cano-Pumarega, I.; et al. Documento internacional de consenso sobre apnea obstructiva del sueño. *Arch. Bronconeumol.* **2022**, *58*, 52–68. [CrossRef]

biomedicines

MDPI

Review

Associated Factors of Pneumonia in Individuals with Chronic Obstructive Pulmonary Disease (COPD) Apart from the Use of Inhaled Corticosteroids

Rosario Lineros [1], Lourdes Fernández-Delgado [2], Antonio Vega-Rioja [2], Pedro Chacón [2], Bouchra Doukkali [2], Javier Monteseirin [3,*] and David Ribas-Pérez [4]

1 Unidad Salud Mental, Hospital Vázquez Díaz, 21080 Huelva, Spain
2 UGC de Alergología, Hospital Universitario Virgen Macarena, 41009 Sevilla, Spain
3 Hospital Quirón Sagrado Corazón, 41013 Sevilla, Spain
4 Department of Stomatology, Faculty of Dentistry, University of Seville, 41004 Seville, Spain
* Correspondence: fmonteseirinmateo@gmail.com

Abstract: Inhaled corticosteroids (ICSs) are widely used in chronic obstructive pulmonary disease (COPD) and in combination with long-acting β2 agonists (LABAs) to reduce exacerbations and improve patient lung function and quality of life. However, ICSs have been associated with an increased risk of pneumonia in individuals with COPD, although the magnitude of this risk remains unclear. Therefore, it is difficult to make informed clinical decisions that balance the benefits and adverse effects of ICSs in people with COPD. There may be other causes of pneumonia in patients with COPD, and these causes are not always considered in studies on the risks of using ICSs in COPD. We consider it very useful to clarify these aspects in assessing the influence of ICSs on the incidence of pneumonia and their role in the treatment of COPD. This issue has important implications for current practice and the evaluation and management of COPD, since COPD patients may benefit from specific ICS-based treatment strategies. Many of the potential causes of pneumonia in patients with COPD can act synergistically, so they can be included in more than one section.

Keywords: COPD; pneumonia; inhaled corticosteroids

1. Introduction

1.1. Pneumonia in Patients with COPD Independent of the Use of ICS

Although pneumonia is often seen as a potential side effect of inhaled corticosteroids (ICSs) in individuals with chronic obstructive pulmonary disease (COPD), the actual frequency of pneumonia varies across studies, and certain analyses have reported varying rates of pneumonia associated with different types of ICS treatments [1–3].

Several randomized controlled trials (RCTs) that involved a significant number of participants did not demonstrate an increase in the occurrence of pneumonia in patients who were randomly assigned to undergo ICS treatment [4–6]. These findings imply that other factors, which are not related to the utilization of ICSs, could impact the probability of developing pneumonia in individuals with COPD.

1.2. Definition of COPD and Pneumonia

The term COPD may not be precise in distinguishing various pulmonary conditions, including asthma, which can pose difficulties in distinguishing asthma from COPD, particularly among smokers and older adults. Moreover, some patients may exhibit clinical features that are common to both asthma and COPD [7]. The diagnosis of pneumonia based solely on clinical signs and laboratory data can be challenging, given the variability of clinical presentations, particularly in the presence of chronic respiratory illness. The literature on COPD often lacks clarity in defining, documenting, classifying, and attributing pneumonia events in terms of severity. Furthermore, variations in the methods used to document and evaluate pneumonia across different countries can contribute to the variability in reported incidence rates, making cross-trial comparisons problematic. Patients with severe forms of COPD may experience exacerbations that can be challenging to differentiate clinically and radiographically from pneumonia, resulting in an overestimation of the number of pneumonia cases in these trials [8].

Patients with acute exacerbations of COPD (AECOPD), a serious public health issue, may exhibit symptoms resembling pneumonia.

Clinically relevant pneumonia can be differentiated from other comparable illnesses using traditional diagnostic criteria including a differential blood count and C-reactive protein (CRP) levels. However, such methods have suboptimal sensitivity and specificity in patients with suspected infections, leading to uncertainty when starting treatment [9].

Bacterial colonization has been linked to increased inflammation, symptom aggravation, and more frequent exacerbations. Nearly 44% of all samples collected from stable COPD patients tested positive for bacterial colonization [10].

2. Various Aspects of Infection in COPD and Its Capacity to Generate Pneumonia

It is widely acknowledged that potentially harmful microorganisms can inhabit the bronchial area in individuals with COPD. Studies have indicated that the composition of the lung microbiome in patients with stable COPD differs significantly from that of healthy individuals [11–18]. There is evidence of a relationship between the appearance of exacerbation symptoms and the acquisition of new bacterial strains [19]. The majority of information about COPD is obtained from samples collected through various methods such as biopsies, lung tissue explants, bronchoalveolar lavage (BAL), protected specimen brush (PSB) techniques, and sputum. Research studies have shown that sputum samples contain different types of microorganisms compared to bronchoalveolar samples, and it has been confirmed that the lungs of COPD patients have unique microbiomes that differ from healthy individuals in terms of bacterial composition [20,21]. Alterations in the microbiome of the upper or lower respiratory tract can affect the immune response and make the host more vulnerable to developing pneumonia [22].

Antibiotic resistance in bacteria that cause community-acquired pneumonia (CAP) such as pneumococci and Mycoplasma pneumoniae is becoming more common, which is a significant global health concern and a contributing factor in the increasing burden of disease, particularly in COPD patients. This information has been reported by several studies, including references [23–25].

Pathogens have several ways of developing resistance to antibiotics, such as producing beta-lactamase, losing outer membrane proteins that are susceptible to antibiotics, changing their targets, forming biofilms, using efflux pumps, and acquiring integrons. The use of antibiotics at low concentrations can also contribute to the development of resistance among the exposed pathogens. Apart from the misuse of antibiotics in healthcare, the general population is exposed to non-iatrogenic antibacterial drugs in their daily lives, such as antibiotics used in the meat industry to treat livestock, leading to an increase in drug resistance in pathogens. The increased use of antibiotics in both clinical and nonclinical settings has resulted in an increasing number of clinically isolated drug-resistant strains, including carbapenem-resistant Klebsiella pneumoniae [26].

The symptoms of SARS-CoV-2 infection can vary widely, from showing no symptoms at all to developing atypical pneumonia. COVID-19 pneumonia is a disease that continues to evolve over time. In fact, around 40% of cases of SARS-CoV-2 infection do not display any symptoms, another 40% have mild upper respiratory symptoms, and approximately 20% develop pneumonia [27–29].

Klebsiella species consist of *Klebsiella ozaenae, Klebsiella rhinoscleroma*, and *Klebsiella pneumoniae,* with the latter having significant clinical implications as an important opportunistic and iatrogenic infectious pathogen. *K. pneumoniae* typically colonize different mucosal surfaces, including the upper respiratory tract, in humans. They are a significant cause of respiratory tract infections that can result in severe pneumonia and are transmitted to the human body via contaminated respirators, atomizers, catheters, and through self-contamination by colonizing bacteria.

The immunological system's growth and maintenance, as well as general health, depend significantly on microflora. Dysbiosis, or changes in the microbial makeup and function in the intestinal and respiratory tract, has recently been connected to variations in immune responses and pulmonary development. Chronic gastrointestinal tract (GIT) disorders such as inflammatory bowel disease (IBD) or irritable bowel syndrome (IBS) frequently coexist with chronic lung diseases such as asthmatic processes or COPD.

Adults with IBD (up to 50%) and IBS (up to 33%) may exhibit symptoms of pulmonary inflammation or compromised pulmonary function. IBD diagnoses are also more common in COPD patients.

In patients with COPD, the intestinal lining is more permeable than in healthy individuals. Although the GIT and respiratory tract have different functions and environments, they share a common embryonic origin and structural similarities. Thus, it is not surprising that they can interact in both health and disease [30–34].

Research indicates that commensal microbes in the respiratory tract may contribute to the development of pneumonia. Furthermore, studies have shown that the intestinal microbiota play a crucial role in regulating local and systemic host responses in clinically relevant models of pneumonia. The intestinal microbiota have been shown to offer protection against pneumonia by priming alveolar macrophages. The interaction between the intestine and lungs during infection has been observed, and it has been suggested that the intestinal microbiota can modulate the immune response in the lungs [35,36].

Periodontal diseases are a group of infectious conditions caused by multiple microorganisms, such as gingivitis and periodontitis, that affect the tissues supporting the teeth. These diseases have been associated with the worsening of various respiratory diseases. The presence of pathogens in the mouth can directly enter the lungs, increasing the risk of acute exacerbations of chronic obstructive pulmonary disease (AECOPD) or community-acquired pneumonia (CAP) in individuals with COPD [37].

Vaccination against influenza, as well as Pneumococcus pneumoniae, has been proven to be effective in preventing infections not only in patients with COPD but also in individuals who are in close contact with them, thus reducing the risk of transmission. Encouraging complete vaccination can contribute to better healthcare management and allocation for patients with COPD [38,39] (Figure 1).

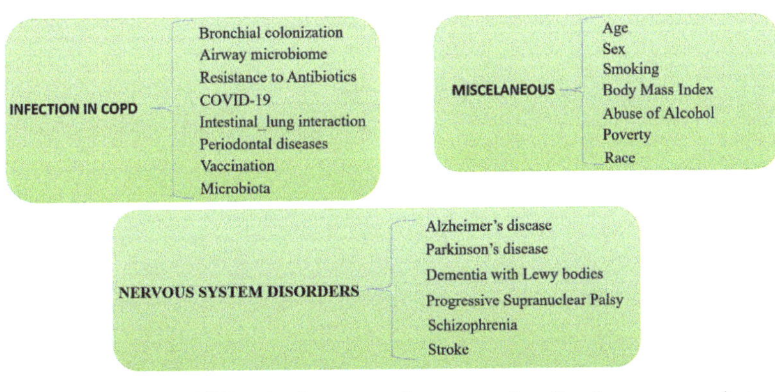

Figure 1. Infections in COPD, miscellaneous, and nervous system disorders as causes of pneumonias in COPD independent of the use of ICS.

2.1. Miscellaneous: Age, Sex, Body Mass Index, Smoking, and Social Status

The elderly are more likely to develop pneumonia due to decreased organ function and an aging immune system, and may suffer multiple organ dysfunction syndrome (MODS), respiratory and circulatory failure, and even death. In fact, pneumonia has become one of the leading causes of death in the elderly [40].

The likelihood of developing community-acquired pneumonia (CAP) depends on age and underlying medical conditions, with older individuals aged 65 or above who have COPD being at a higher risk. Within the COPD population, those aged 65–79 or over 80 are more susceptible to CAP than those aged 45–65. Several independent factors have been identified for recurrent CAP in adults, including COPD, older age, and a lack of pneumococcal vaccination [6,40–44].

The impact of seasonal influenza on health varies based on age and pre-existing medical conditions. Individuals aged 65 or older with COPD are at higher risk of serious medical complications leading to hospitalization and death. COPD is the third leading cause of death among people aged 65 and above in the United States. About 15 million adults in the US have been diagnosed with COPD, and 5 million of them are aged 65 or older. For people with COPD, influenza infection can cause severe complications and even death, especially in older adults. Patients with COPD have a higher risk of respiratory failure and are more likely to exhibit frailty. Any respiratory infection, including influenza, can cause airway inflammation and constriction, making it difficult to breathe properly. This inflammation can increase the likelihood of COPD exacerbations. As a result, people with COPD are advised to receive seasonal influenza vaccination [45].

Adjusted incidences of all types of pneumonia, including community-acquired pneumonia (CAP) and hospital-acquired pneumonia (HAP), are higher in men than in women. Specifically, men have a 47% higher adjusted incidence of CAP and a 98% higher adjusted incidence of HAP than women. Furthermore, men with COPD have a higher risk of developing pneumonia than women with COPD [46].

A single predictive factor for COPD is body mass index (BMI), and there is a direct correlation between lower body weight and higher death rates in COPD patients. While evidence suggest that people who can acquire weight have a better prognosis, more weight loss increases the danger. The prognostic relevance of a low BMI could be the result of waste in end-stage COPD. In turn, underweight patients may be less resistant to infections such as pneumonia [47–49].

Additionally, smoking has been recognized as a risk factor for CAP, with the likelihood altering depending on the history of smoking. Smoking's impact on the risk of CAP has been studied. Smoking was verified as an independent risk factor in the multivariate analysis, and both current smokers and former smokers were shown to have a greater risk of CAP compared to non-smokers. With higher pack years, the risk grew higher. The risk of CAP was substantially greater in ex-smokers who had quit smoking more recently (4 years) compared to those who had quit smoking less recently. Those who had never smoked but were exposed to passive smoking were similarly at a significantly higher risk of CAP. Likewise, elderly individuals aged 65 years or older who were subjected to involuntary exposure to passive smoke within their domestic environment exhibited a heightened susceptibility to community-acquired pneumonia (CAP). It has been observed that individuals who are current smokers and aged 65 years or above, are at a greater risk of contracting the infection compared to those who are ex-smokers, regardless of their age [4,6,50–55].

According to the Centers for Disease Control and Prevention (CDC), people who abuse alcohol are 10 times more likely to develop pneumococcal pneumonia and 4 times more likely to die from pneumonia than non-drinkers [56,57]. Heavy drinkers are more likely to suffer from COPD according to the Global Initiative for Chronic Obstructive Lung Disease (GOLD) criteria and suffer more respiratory symptoms [58].

Poverty is associated with an increased risk of pneumonia [59]. Although race is a characteristic that must be taken into account, it can be conditioned by other co-existing factors such as socioeconomic condition. For example, Black adults in the United States are 2.6 times more likely to develop pneumonia than Caucasian adults, although in general, the former are also 4.4 times poorer than Caucasian adults [60,61] (Figure 1).

2.2. Nervous System Disorders

Advanced-stage Alzheimer's disease (AD) is often accompanied by two serious medical conditions: dysphagia and aspiration pneumonia. Dysphagia associated with pseudobulbar palsy can cause weight loss, despite attempts to manage swallowing difficulties. The underlying mechanisms that regulate the body's basic functions, such as maintaining nutritional status, appear to be impaired in these patients. Aspiration pneumonia is the leading cause of death in end-stage AD.

Aspiration pneumonia is caused by a number of factors such as dysphagia, reduced consciousness, a loss of gag reflex, periodontal disease, and the mechanical effects of inserting tubes into the respiratory and gastrointestinal tracts. The bacteria responsible for the infection include oral flora, mostly aerobes, and nosocomial-acquired pathogens, such as Staphylococcus aureus, aerobic and facultative Gram-negative bacilli, which may colonize patients in hospitals and nursing homes. Besides the use of antibiotics, addressing the symptoms and managing AD in patients with pneumonia is critical in alleviating their suffering [32].

Neurological disorders such as Alzheimer's disease, Parkinson's disease, Lewy bodies, and progressive supranuclear palsy have been found to increase the risk of pneumonia that is confirmed through autopsy. Patients with these conditions are often bedridden and may experience dysphagia, altered mental states, or respiratory muscle weakness, all of which can increase the risk of pneumonia. Additionally, the causes of pneumonia in patients with neurological conditions who also have COPD may be different from those in the general population, which can lead to differences in the distribution of pathogens causing the infection [62].

Patients with schizophrenia are at increased risk of developing pneumonia, as well as other respiratory diseases such as COPD and asthma [63].

HAD is one of the most important complications that can be prevented in a patient who has suffered a stroke. The occurrence of pneumonia worsens the prognosis of a patient, especially if he or she has had an ischemic stroke, reducing the chances of a full recovery and increasing the chances of bodily deterioration that can lead to the patient's death. Although the chances of developing pneumonia in the general population range from approximately 8% to 30%, it increases to 44% in patients with COPD and who have had a stroke [64] (Figure 1)

3. Hospitalization and Similar Conditions

Hospital-acquired pneumonia (HAP) is a type of infection that patients acquire during their hospital stay. It is a common nosocomial infection that causes significant clinical and economic burdens, including prolonged hospitalization, high medical costs, and increased morbidity and mortality.

Hospitalization alone can increase the risk of pneumonia, but there are additional factors associated with hospitalization that can further increase this risk. These factors include older age, male gender, pre-existing asthma, pre-existing COPD, pre-existing chronic lower airway disease, tube feeding, suctioning, positioning, use of mechanical ventilation, admission to the ICU, poverty, race, anemia, dementia, chronic kidney disease, paraplegia, hemiplegia, and metastatic carcinoma. These factors may differ from those seen in the general population and may increase the risk of pneumonia in hospitalized patients.

- Factors specific to hospitalization.
- Type of hospital: It has been reported that the incidence of hospital-acquired pneumonia (HAP) in tertiary hospitals is lower than in general hospitals. Specifically, the incidence in tertiary hospitals is around 3.5%, while in general hospitals, it is around 5.7%. This difference could be attributed to variations in the quality of patient care and access to healthcare resources between the two types of hospitals. Tertiary hospitals may have better environmental hygiene practices and more highly trained healthcare professionals, which could lead to better quality care and ultimately a lower incidence of HAP.

- Type of patient: The prevalence of HAP is relatively higher among medical patients compared to surgical patients. Nevertheless, it is important to acknowledge that surgical patients remain susceptible to the onset of HAP.
- Higher bed-to-nurse ratios: Patients who are hospitalized in facilities that exhibit bed-to-nurse ratios rated as g 4 and 5 experience a 1.4-fold elevated risk of HAP in comparison to those under the care of hospitals that maintain a bed-to-nurse ratio classified as grade 1. From a practical standpoint, nurses who are assigned a reduced patient load would be in a position to allocate more time and resources towards the provision of care to those patients who are entrusted under their purview.
- Type of room: The incidence of HAP exhibits a notable approximately threefold increase in patients whose accommodation comprises more than four beds, compared to those who are placed in rooms containing three or less beds. Patients who are accommodated in individual patient rooms are likely to experience a reduced risk of HAP compared to their counterparts residing in multi-patient rooms. The mitigation of HAP risk can be attributed to the limited exposure of patients in single-patient rooms to potential reservoirs, and attributed to limited or no direct and indirect contacts [58–63].

Aspiration pneumonia (AP) is an infectious condition caused by inhalation of oropharyngeal secretions populated by pathogenic bacteria, whereas aspiration pneumonitis (Mendelson syndrome) (MS) is a chemical injury brought on by the inhalation of sterile gastric contents.

Factors predisposing to MS and AP include a decreased level of consciousness, neurological disorders, dysphagia, and material aspiration in association with tracheostomy [37,38] (Figure 2).

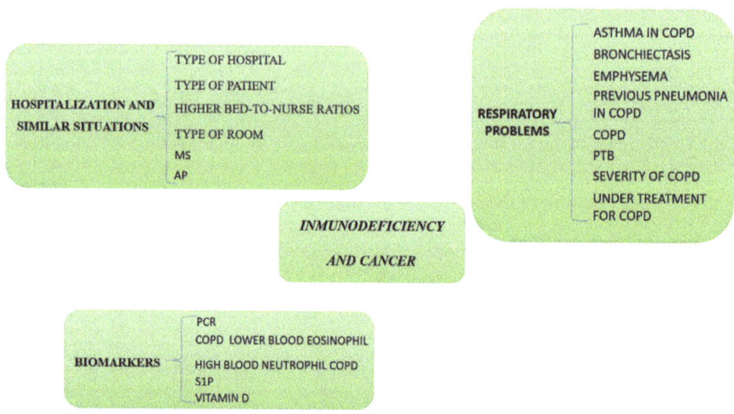

Figure 2. Hospitalization and similar situations, respiratory problems, immunodeficiencies, and cancer as causes of pneumonias in COPD independent of ICS use. PCR: C-reactive protein. S1P: Sphingosine-1-phosphate. NLR: neutrophil–lymphocyte ratio. AP: Aspiration pneumonia. MS: Mendelson syndrome. PTB: Pulmonary tuberculosis.

3.1. Respiratory Problems

The presence of asthma has been identified as an independent risk factor for pneumonia in the population of COPD and has been associated with a 13% risk of pneumonia [39]. Asthma has been shown to be the strongest independent risk factor for pneumonia in patients with COPD, resulting in more exacerbations and more severe limitation of airflow compared with patients with COPD alone [16].

Structural changes in COPD, such as bronchiectasis, can modulate the severity of the exacerbation and contribute to the morbidity associated with the latter. In 2014, GOLD first described bronchiectasis as one of the comorbidities of COPD [42].

Bronchiectasis and COPD are found to coexist in 20–60% of all cases. However, patients with severe COPD are reported to have a particularly high frequency of bronchiectasis, although the published figures vary widely between 4 and 72%. In patients with COPD and bronchiectasis, the main clinical manifestations are those typically associated with bronchiectasis: chronic cough and sputum production, chronic bronchial infection, and frequent infective exacerbations [43,44].

Computed tomography (CT)-diagnosed emphysema has been identified as the strongest risk factor for pneumonia among all clinical parameters in patients with COPD. The presence of emphysema has been associated with the severity of pneumonia in patients with COPD. Lower lobe emphysema is particularly associated with the frequency of exacerbation in COPD, according to the involvement of lung structures such as diameters [52,53].

An important factor that can be a predisposition to pneumonia in COPD is a history of previous pneumonia episodes. Patients with COPD who have suffered pneumonia before have an increase incidence of CAP [6,16].

Patients with previous COPD are at increased risk of developing pneumonia compared to otherwise healthy individuals and often have poorer clinical outcomes in terms of the severity of the pneumonia [16].

Mycobacterium tuberculosis has the ability to infiltrate a multitude of body organs. Pulmonary tuberculosis (PTB) frequently manifests as the predominant clinical form of the disease, primarily characterized by pulmonary involvement resulting in respiratory impairment. This ailment is a common chronic consumptive illness; however, it may manifest itself as an acute pneumonia. Acute tuberculous pneumonia (TP) bears resemblance to conventional bacterial pneumonia in its clinical presentation. Acute TP is commonly associated with the clinical presentations of CAP; however, in the present scenario, the etiological agent is microplasma. Tuberculosis should be preferred over non-tuberculous bacteria or viruses. In developing nations, *M. Tuberculosis* serves as the primary etiological agent responsible for the onset of community-acquired pneumonia (CAP). On the other hand, the relationship between tuberculosis and COPD is well known [65,66].

Several studies have shown that an increase in infections, especially respiratory infections, increases the risk of developing pneumonia [36,67,68].

The co-occurrence of inflammatory diseases in both the upper and lower respiratory tracts is a common phenomenon. Sino-pulmonary disease is a widely acknowledged medical condition, particularly in cases where the ailment has progressed to a chronic stage. Individuals who exhibit significant sphenoid involvement, characterized by either complete or partial opacification, possess a heightened likelihood of being diagnosed with CAP, with a relative risk of 20 times that of individuals without such involvement [69].

Significant differences in the risk of pneumonia have been reported according to the severity of COPD. The severity of the underlying respiratory disease affects the risk of CAP. Studies on individuals aged 65 years and above with mild lung disease, i.e., not necessitating medication or oxygen, have indicated a twofold increase in the likelihood of contracting community-acquired pneumonia compared to those without lung disease. Conversely, individuals with severe lung disease demanding oxygen therapy exhibit an eightfold greater probability of developing community-acquired pneumonia. The presence of moderate and severe lung disease, as indicated by a predicted percentage forced expiratory volume in one second (FEV1) of 50–80%, has been recognized as a noteworthy risk factor for community-acquired pneumonia (CAP) in older adults aged 65 years and above, in contrast to those possessing healthy or mildly compromised pulmonary function. Moderate COPD exacerbation and hospitalization due to severe COPD exacerbation have also been identified as independent risk factors for CAP in patients with COPD $\geq$ 45 years of age. The inclusion of patients with these characteristics in study populations is very likely to contribute to the variation in reported pneumonia rates between studies [2,6,8,16,55,70–73].

Despite the prevalence of evidence-based guidelines pertaining to COPD management, findings from the survey indicate that healthcare professionals display notable deficiencies in their comprehension of essential aspects inherent to the management of this condition. Specifically, at least 50% of surveyed practitioners reported being unaware of established guidelines for both diagnosing and treating COPD [11]. Insufficient awareness of recommended treatment modalities can potentially result in suboptimal management of individuals diagnosed with COPD. Patients receiving COPD treatment are more likely to have more comorbidities, including pneumonia [74] (Figure 2).

3.2. Immunodeficiency and Cancer

Patients with compromised immunity exhibit a heightened susceptibility to contracting CAP in comparison to the general populace. This demographic displays an impaired immune system, thereby resulting in a diminished resilience to pathogenic agents. The innate condition of immunosuppression is one possible etiology; nonetheless, the acquired form of immunodeficiency is considerably more prevalent owing to the progression of cancer chemotherapy in recent years. In essence, a strong correlation exists between chronic inflammation and impaired immune response to respiratory pathogens, which promotes the activation of immunosuppressive pathways. This process is believed to be a contributing factor in the pathogenesis of COPD. It is crucial to bear in mind that an immunocompromised individual may harbor multiple concomitant infections.

Some of the following circumstances may contribute to a compromised immune system:

Hematologic malignancies, with the use of corticosteroids and monoclonal antibodies, and T cell dysfunction.

Solid tumors, with high-dose chemotherapy administration, prolonged use of corticosteroids, and marrow transplantation.

Solid organ transplantation, with decreased CD4+ cell counts.

Autoimmune conditions, particularly among individuals receiving immunomodulatory medications such as anti-TNF-α, present a significant clinical concern. Individuals are administered glucocorticoids, as well as other biological and immunomodulatory therapeutic agents for treatment of various medical conditions [75–77].

Complications pertaining to human immunodeficiency virus (HIV) primarily affect the lungs, rendering them a crucial target. Observably, individuals with HIV infection are predisposed to a myriad of opportunistic pneumonias, neoplasms, and pulmonary disorders. Opportunistic pneumonias are recognized as the primary causes of morbidity and mortality within the context of pulmonary complications associated with human immunodeficiency virus (HIV). The spectrum of opportunistic pneumonias associated with HIV encompasses a diverse array of pathogens, encompassing bacteria, mycobacteria, fungi, viruses, and parasites.

The prevalence of COPD is higher in people with HIV compared to those without HIV infection. The higher prevalence is explained in part by the higher incidence of smoking in people living with HIV and may be related to an increased susceptibility to respiratory infections, immunosuppressed states, and chronic inflammation [78–81].

Any type of cancer can directly or indirectly (chemotherapy, hospitalization, metastasis, etc.) lead to pneumonia in COPD, especially in patients with lung cancer or subjects that have been operated upon due to lung cancer. In fact, lung cancer is an important comorbid condition in patients with COPD. The prevalence of lung cancer in COPD is reported to be 2.79%. People with COPD were found to be 6.35 times more likely to have lung cancer than controls. Reduced survival has been reported in patients with lung cancer with comorbidity in the form of COPD [45,46] (Figure 2).

3.3. Biomarkers

C-reactive protein (CRP) has been widely used in the management of pneumonia. It is a well-established biomarker of inflammation, but has been regarded as a nonspecific marker in the diagnosis of pneumonia. Nevertheless, it may be of some use in defining the severity of pneumonia [49,50].

Taking into account a baseline eosinophil count of 2% as the threshold value, patients with COPD with lower blood eosinophil counts suffer more pneumonia events than those with higher counts [54].

High blood neutrophil counts in COPD have been associated with an increased risk of pneumonia, independent of ICS use [82].

The neutrophil/lymphocyte ratio (NLR) has been reported to be related to mortality and prognosis in patients with CAP, with better performance as a marker than CRP. The NLR, upon admission to the emergency department, predicts the severity and outcome of CAP with greater prognostic precision than traditional infection markers [83–85].

Sphingosine-1-phosphate (S1P) is a bioactive sphingolipid that participates in numerous physiological processes, such as immune responses and maintenance of the endothelial barrier's integrity. Moreover, sphingosine-1-phosphate (S1P) exhibits potential as a diagnostic and prognostic biomarker for the preliminary evaluation of individuals presenting with pneumonia. The present study indicates a remarkably substantial elevation in the levels of plasma sphingosine-1-phosphate (S1P) in individuals with pneumonia, which also display a positive correlation with the severity of the disease [9,86].

Serum levels of 25-hydroxyvitamin D [25(OH)D] have been shown to be associated with the risk of CAP. Those subjects with 25(OH)D levels < 30 ng/mL had a significantly higher risk of CAP compared to those with levels ≥ 30 ng/mL [87] (Figure 2).

4. Chronic Medical Conditions

Compared to patients without COPD, patients with pneumonia with COPD are likely to have more severe pneumonia, with an increase in hospital admissions, and a poorer outcome [18–20]. In the first year after the diagnosis of COPD, people are at a 16-fold higher risk of pneumonia compared to patients without COPD. The incidence rate of CAP was found to be 22.4 events per 1000 person-years in the 10 years after the diagnosis of COPD, and more than 50% higher among those classified as having severe COPD [6]. The presence of three or more chronic medical conditions, commonly referred to as multiple morbidity, has been found to be significantly associated with an elevated risk of pneumococcal disease in patients in the age range of 50–64 years. Furthermore, a positive correlation has been established between augmented occurrences of persistent medical conditions and an elevated prevalence of pneumococcal ailment in individuals aged 65 years or above [51].

Individuals afflicted with chronic heart disease (CHD), encompassing congestive heart failure (CHF) and cardiovascular and valve maladies, are vulnerable to contracting CAP with an up to 3.3-fold elevated risk as opposed to those without CHD. Such a risk is contingent upon both the specific condition and advancing age of the afflicted individual. Among individuals aged 65 years or older in the United States, heart disease has been established as a distinct risk factor for CAP, with approximately 16% of total CAP cases being linked to the aforementioned underlying condition.

The risk of CAP was found to be higher in patients with more severe heart disease, while individuals with non-CHF heart disease showed only a modest increase in the risk of CAP compared to those without heart disease. Patients with pneumonia and CHF were found to have an almost two-fold increased risk of hospitalization due to pneumonia compared to those without CHF. The susceptibility to pneumonia was found to be contingent upon both the comorbidities accompanying heart failure and the nature of the administered medical intervention. Individuals diagnosed with cardiomyopathy, as well as those receiving loop diuretic therapy, exhibit a heightened susceptibility to hospitalization resulting from CAP. Amiodarone has been identified as a distinct risk factor for the development of CAP among individuals undergoing treatment for heart failure. Various studies have reported this association, highlighting the need to exercise caution when administering this medication to such patients [16].

Patients with diabetes have an up to 1.4 times higher risk of developing CAP, and the global risk of pneumonia is 1.4-to-4.6-fold higher than in non-diabetic individuals [16].

The prevalence of gastro-esophageal reflux disease (GERD) in individuals with COPD ranges from 17 to 54%. The presence of GERD has been shown to be associated with a higher risk of AECOPD and pneumonia. This is consistent with a defined phenotype for patients with COPD who experience frequent AECOPD (two per year), and with GERD being identified as an independent predictor [55–57].

Pneumococcal disease is often associated with the presence of numerous risk factors in individuals aged 65 years or older, with an estimated 60% exhibiting two or more underlying medical conditions. Numerous conditions have demonstrated a cumulative impact on both the susceptibility of contracting CAP and the associated fatality rate of this pathology. The prevalence of pneumococcal disease escalates proportionately with the amplification of risk factors. The incidence rates are notably elevated for individuals with three or more predisposing conditions [70,88] (Figure 3).

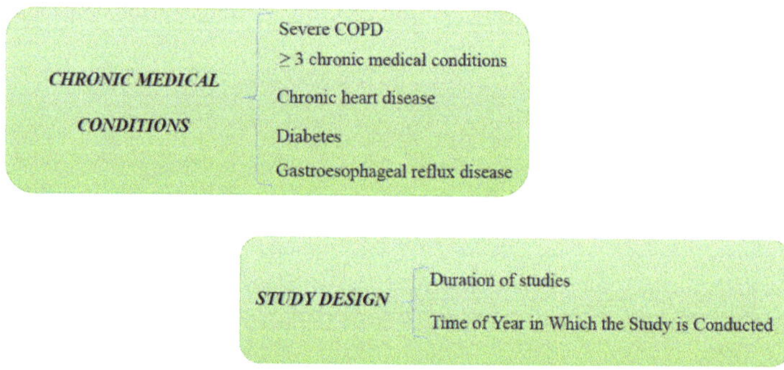

Figure 3. Chronic medical conditions and study design as causes of pneumonia in COPD independent of ICS use.

4.1. Study Design

The study design can be of great importance in terms of the final results obtained. Furthermore, the duration of studies can be highly variable and, in this regard, the longer the duration, the more likely it is that a pneumonic process develops [89].

The time of year in which the study is conducted can also exert an influence, with more cases of pneumonia recorded in studies that are carried out in winter compared to studies conducted in summer [70].

Elements such as the design of the period prior to starting study treatment are likely to affect the risk that patients develop pneumonia [90].

The absence of randomization in observational studies leads to the risk of confusion by indication [8] (Figure 3).

4.2. Triple Therapy

Although the use of triple therapy in COPD appears to have several beneficial parameters, with regard to the risk it represents for these patients due to the possible occurrence of pneumonias, the different studies make the same mistakes described in this review [91,92].

5. Conclusions

An increased risk of pneumonia with inhaled corticosteroids in COPD was first described in 2007, and has since been reported in both randomized trials and observational studies. Despite the number of studies reporting this association, many unresolved issues remain regarding the relationship between inhaled corticosteroid use and pneumonia in individuals with COPD.

The idea that the incidence of pneumonia increases in COPD patients treated with ICS, derived from both observational studies and randomized controlled trials, is not supported by firm data, however.

As there is a possibility that the use of ICS in COPD may increase the risk of pneumonia, the elements considered here must be taken into account before prescribing ICS, because they may increase this risk (Figure 4).

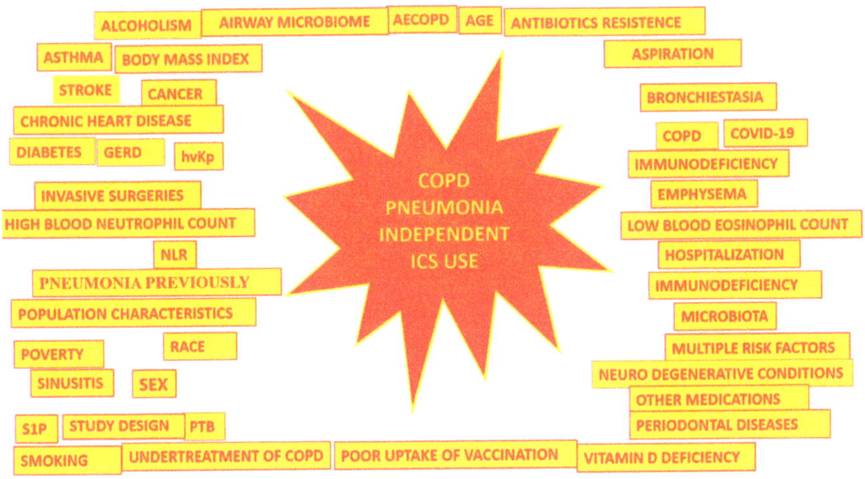

Figure 4. Causes of pneumonia in COPD that are combined and unrelated to ICS usage.

Author Contributions: Conceptualization, J.M.; methodology, R.L. and L.F.-D.; software, D.R.-P.; validation, A.V.-R. and P.C.; formal analysis, J.M.; investigation, J.M. and B.D.; resources, D.R.-P.; data curation, J.M.; writing—original draft preparation, J.M.; writing—review and editing, J.M. and D.R.-P.; supervision, J.M. All authors have read and agreed to the published version of the manuscript.

Funding: This work was co-funded by the European Union through the European Regional Development Fund (ERDF) and was supported by grants from the Instituto de Salud Carlos III (FIS-Thematic Networks and Cooperative Research Centers ARADYAL, RD16/0006/0035) and Fundación Alergol, Spain. A.V.-R. was supported by a grant from the Ministerio de Economía y Competitividad (Proyectos I+D+i para Jóvenes Investigadores, SAF2014-60649-JIN) and holds a Nicolás Monardes contract from the Andalusian Health Service (C-0060-2018); P.C. is under a senior postdoc contract from the Ministry of Health and Families (Junta de Andalucía, Ref RH-0129-2020). The funders had no role in the study design, the data collection and analysis, the decision to publish, or the preparation of the manuscript.

Institutional Review Board Statement: Not aplicable.

Informed Consent Statement: Not applicable.

Data Availability Statement: The data presented in this study are available on request from the corresponding author.

Conflicts of Interest: The authors declare no conflict of interest. The funders had no role in the design of the study; in the collection, analyses, or interpretation of the data; in the writing of the manuscript; or in the decision to publish the results.

References

1. Zhang, Q.; Li, S.; Zhou, W.; Li, J.; Cao, J. Risk of Pneumonia with Different Inhaled Corticosteroids in COPD Patients: A Meta-Analysis. *COPD* **2020**, *17*, 462–469. [CrossRef] [PubMed]
2. Chen, H.; Sun, J.; Huang, Q.; Yuan, M.; Ma, C.; Yan, H. Inhaled Corticosteroids and the Pneumonia Risk in Patients with Chronic Obstructive Pulmonary Disease: A Meta-analysis of Randomized Controlled Trials. *Front. Pharmacol.* **2021**, *12*, 691621. [CrossRef] [PubMed]
3. Miravitlles, M.; Auladell-Rispau, A.; Monteagudo, M.; Vázquez-Niebla, J.C.; Mohammed, J.; Nuñez, A.; Urrútia, G. Systematic review on long-term adverse effects of inhaled cor-ticosteroids in the treatment of COPD. *Eur. Respir. Rev.* **2021**, *30*, 210075. [CrossRef]

4. Almirall, J.; Serra-Prat, M.; Bolíbar, I.; Balasso, V. Risk Factors for Community-Acquired Pneumonia in Adults: A Systematic Review of Observational Studies. *Respiration* **2017**, *94*, 299–311. [CrossRef]
5. Jackson, M.L.; Neuzil, K.M.; Thompson, W.W.; Shay, D.; Yu, O.; Hanson, C.A.; Jackson, L.A. The burden of community-acquired pneumonia in seniors: Results of a population-based study. *Clin. Infect. Dis.* **2004**, *39*, 1642–1650. [CrossRef]
6. Müllerova, H.; Chigbo, C.; Hagan, G.W.; Woodhead, M.A.; Miravitlles, M.; Davis, K.J.; Wedzicha, J.A. The natural history of community-acquired pneumonia in COPD patients: A population database analysis. *Respir. Med.* **2012**, *106*, 1124–1133. [CrossRef]
7. von Wichert, P. Bronchial Diseases are Insufficiently Defined with the Term COPD. *Int. J. Chron. Obstruct. Pulmon. Dis.* **2021**, *16*, 1349–1352. [CrossRef]
8. Wise, R.A.; Bafadhel, M.; Crim, C.; Criner, G.J.; Day, N.C.; Halpin, D.M.G.; Han, M.K.; Lange, P.; Lipson, D.A.; Martinez, F.J.; et al. Discordant diagnostic criteria for pneumonia in COPD trials: A review. *Eur. Respir. Rev.* **2021**, *30*, 210124. [CrossRef]
9. Hsu, C.-W.; Suk, C.-W.; Hsu, Y.-P.; Chang, J.-H.; Liu, C.-T.; Huang, S.-K.; Hsu, S.-C. Sphingosine-1-phosphate and CRP as potential combination biomarkers in discrimination of COPD with community-acquired pneumonia and acute exacerbation of COPD. *Respir. Res.* **2022**, *23*, 63. [CrossRef] [PubMed]
10. Armitage, M.N.; Spittle, D.A.; Turner, A.M. A Systematic Review and Meta-Analysis of the Prevalence and Impact of Pulmonary Bacterial Colonisation in Stable State Chronic Obstructive Pulmonary Disease (COPD). *Biomedicines* **2021**, *10*, 81. [CrossRef]
11. Chen, B.; Liu, W.; Chen, Y.; She, Q.; Li, M.; Zhao, H.; Zhao, W.; Peng, Z.; Wu, J. Effect of Poor Nutritional Status and Comorbidities on the Occurrence and Outcome of Pneumonia in Elderly Adults. *Front. Med.* **2021**, *8*, 719530. [CrossRef]
12. Garcia-Vidal, C.; Carratalà, J.; Fernández-Sabé, N.; Dorca, J.; Verdaguer, R.; Manresa, F.; Gudiol, F. Aetiology of, and risk factors for, recurrent community-acquired pneumonia. *Clin. Microbiol. Infect.* **2009**, *15*, 1033–1038. [CrossRef]
13. Gau, J.T.; Acharya, U.; Khan, S.; Heh, V.; Mody, L.; Kao, T.C. Pharmacotherapy and the risk for community-acquired pneumonia. *BMC Geriatr.* **2010**, *10*, 45. [CrossRef] [PubMed]
14. Jackson, M.L.; Nelson, J.C.; Jackson, L.A. Risk factors for community-acquired pneumonia in immunocompetent seniors. *J. Am. Geriatr. Soc.* **2009**, *57*, 882–888. [CrossRef]
15. Loeb, M.; Neupane, B.; Walter, S.D.; Hanning, R.; Carusone, S.C.; Lewis, D.; Krueger, P.; Simor, A.E.; Nicolle, L.; Marrie, T.J. Environmental risk factors for community-acquired pneumonia hospitalization in older adults. *J. Am. Geriatr. Soc.* **2009**, *57*, 1036–1040. [CrossRef]
16. Torres, A.; Blasi, F.; Dartois, N.; Akova, M. Which individuals are at increased risk of pneumococcal disease and why? Impact of COPD, asthma, smoking, diabetes, and/or chronic heart disease on community-acquired pneumonia and invasive pneumococcal disease. *Thorax* **2015**, *70*, 984–989. [CrossRef]
17. Young-Xu, Y.; Smith, J.; Nealon, J.; Mahmud, S.M.; Van Aalst, R.; Thommes, E.W.; Neupane, N.; Lee, J.K.H.; Chit, A. Influenza vaccine in chronic obstructive pulmonary disease among elderly male veterans. *PLoS ONE* **2022**, *17*, e0262072. [CrossRef] [PubMed]
18. Erb-Downward, J.R.; Thompson, D.L.; Han, M.K.; Freeman, C.M.; McCloskey, L.; Schmidt, L.A.; Young, V.B.; Toews, G.B.; Curtis, J.L.; Sundaram, B.; et al. Analysis of the lung microbiome in the "healthy" smoker and in COPD. *PLoS ONE* **2011**, *6*, e16384. [CrossRef] [PubMed]
19. Sze, M.A.; Dimitriu, P.A.; Hayashi, S.; Elliott, W.M.; McDonough, J.E.; Gosselink, J.V.; Cooper, J.; Sin, D.D.; Mohn, W.W.; Hogg, J.C. The lung tissue microbiome in chronic obstructive pulmonary disease. *Am. J. Respir. Crit. Care Med.* **2012**, *185*, 1073–1080. [CrossRef]
20. Pragman, A.A.; Kim, H.B.; Reilly, C.S.; Wendt, C.; Isaacson, R.E. The lung microbiome in moderate and severe chronic obstructive pulmonary disease. *PLoS ONE* **2012**, *7*, e47305. [CrossRef] [PubMed]
21. Huang, Y.J.; Kim, E.; Cox, M.J.; Brodie, E.; Brown, R.; Wiener-Kronish, J.P.; Lynch, S.V. A persistent and diverse airway microbiota present during chronic obstructive pulmonary disease exacerbations. *OMICS* **2010**, *14*, 9–59. [CrossRef]
22. Rogers, G.B.; Daniels, T.W.; Tuck, A.; Carroll, M.P.; Connett, G.J.; David, G.J.; Bruce, K.D. Studying bacteria in respiratory specimens by using conventional and molecular microbiological approaches. *BMC Pulm. Med.* **2009**, *9*, 14. [CrossRef]
23. Millares, L.; Ferrari, R.; Gallego, M.; Garcia-Nuñez, M.; Brocal, V.P.; Espasa, M.; Pomares, X.; Monton, C.; Moya, A.; Monsó, E. Bronchial microbiome of severe COPD patients colonised by Pseudomonas aeru-ginosa. *Eur. J. Clin. Microbiol. Infect. Dis.* **2014**, *33*, 1101–1111. [CrossRef] [PubMed]
24. Garcia-Nuñez, M.; Millares, L.; Pomares, X.; Ferrari, R.; Pérez-Brocal, V.; Gallego, M.; Espasa, M.; Moya, A.; Monsó, E. Severity-related changes of bronchial microbiome in chronic obstructive pulmonary disease. *J. Clin. Microbiol.* **2014**, *52*, 4217–4223. [CrossRef]
25. Cabrera-Rubio, R.; Garcia-Núñez, M.; Setó, L.; Antó, J.M.; Moya, A.; Monsó, E.; Mira, A. Microbiome diversity in the bronchial tracts of patients with chronic obstructive pulmonary disease. *J. Clin. Microbiol.* **2012**, *50*, 3562–3568. [CrossRef]
26. Sethi, S.; Murphy, T.F. Infection in the pathogenesis and course of chronic obstructive pulmonary disease. *N. Engl. J. Med.* **2008**, *359*, 2355–2365. [CrossRef]
27. Dickson, R.P.; Erb-Downward, J.R.; Martinez, F.J.; Huffnagle, G.B. The Microbiome and the Respiratory Tract. *Annu. Rev. Physiol.* **2016**, *78*, 481–504. [CrossRef]
28. Wu, B.G.; Segal, L.N. The Lung Microbiome and Its Role in Pneumonia. *Clin. Chest Med.* **2018**, *39*, 677–689. [CrossRef]
29. Luján, M.; Gallego, M.; Belmonte, Y.; Fontanals, D.; Vallés, J.; Lisboa, T.; Rello, J. Influence of pneumococcal serotype group on outcome in adults with bacteraemic pneumonia. *Eur. Respir. J.* **2010**, *36*, 1073–1079. [CrossRef]

30. Simet, S.M.; Sisson, J.H. Alcohol's Effects on Lung Health and Immunity. *Alcohol. Res.* **2015**, *37*, 199–208.
31. Frantz, S.; Wollmer, P.; Dencker, M.; Engström, G.; Nihlén, U. Associations between lung function and alcohol consumption–assessed by both a questionnaire and a blood marker. *Respir. Med.* **2014**, *108*, 114–121. [CrossRef] [PubMed]
32. Kalia, M. Dysphagia and aspiration pneumonia in patients with Alzheimer's disease. *Metabolism* **2003**, *52* (Suppl. S2), 36–38. [CrossRef] [PubMed]
33. Giske, C.G.; Monnet, D.L.; Cars, O.; Carmeli, Y. ReAct-Action on Antibiotic Resistance. Clinical and economic impact of common multidrug-resistant gram-negative bacilli. *Antimicrob. Agents Chemother.* **2008**, *52*, 813–821. [CrossRef] [PubMed]
34. Pendleton, J.N.; Gorman, S.P.; Gilmore, B.F. Clinical relevance of the ESKAPE pathogens. *Expert Rev. Anti. Infect. Ther.* **2013**, *11*, 297–308. [CrossRef]
35. Morimoto, K.; Suzuki, M.; Ishifuji, T.; Yaegashi, M.; Asoh, N.; Hamashige, N.; Abe, M.; Aoshima, M.; Ariyoshi, K.; Adult Pneumonia Study Group-Japan (APSG-J). The burden and etiology of community-onset pneumonia in the aging Japanese population: A multicenter prospective study. *PLoS ONE* **2015**, *10*, e0122247. [CrossRef]
36. Chang, D.; Sharma, L.; Dela Cruz, C.S.; Zhang, D. Clinical Epidemiology, Risk Factors, and Control Strategies of Klebsiella pneumoniae Infection. *Front. Microbiol.* **2021**, *12*, 750662. [CrossRef]
37. Finegold, S.M. Aspiration pneumonia. *Rev. Infect. Dis.* **1991**, *13* (Suppl. S9), S737–S742. [CrossRef]
38. Marik, P.E. Aspiration pneumonitis and aspiration pneumonia. *N. Engl. J. Med.* **2001**, *344*, 665–671. [CrossRef]
39. Janson, C.; Johansson, G.; Ställberg, B.; Lisspers, K.; Olsson, P.; Keininger, D.L.; Uhde, M.; Gutzwiller, F.S.; Jörgensen, L.; Larsson, K. Identifying the associated risks of pneumonia in COPD patients: ARCTIC an observational study. *Respir. Res.* **2018**, *19*, 172. [CrossRef]
40. Wilson, D.O.; Rogers, R.M.; Wright, E.C.; Anthonisen, N.R. Body weight in chronic obstructive pulmonary disease. The National Institutes of Health Intermittent Positive-Pressure Breathing Trial. *Am. Rev. Respir. Dis.* **1989**, *139*, 1435–1438. [CrossRef] [PubMed]
41. Schols, A.M.; Slangen, J.; Volovics, L.; Wouters, E.F. Weight loss is a reversible factor in the prognosis of chronic obstructive pulmonary disease. *Am. J. Respir. Crit. Care Med.* **1998**, *157 Pt 1*, 1791–1797. [CrossRef] [PubMed]
42. Global Strategy for the Diagnosis, Management, and Prevention of Chronic Obstructive Pulmonary Disease (updated 2014) [EB/OL]. Available online: http://www.goldcopd.org/GlobalInitiativeforChronicObstructiveLungDisease (accessed on 5 December 2022).
43. Polverino, E.; Dimakou, K.; Hurst, J.; Martinez-Garcia, M.-A.; Miravitlles, M.; Paggiaro, P.; Shteinberg, M.; Aliberti, S.; Chalmers, J.D. The overlap between bronchiectasis and chronic airway diseases: State of the art and future directions. *Eur. Respir. J.* **2018**, *52*, 1800328. [CrossRef] [PubMed]
44. Ni, Y.; Shi, G.; Yu, Y.; Hao, J.; Chen, T.; Song, H. Clinical characteristics of patients with chronic obstructive pulmonary disease with comorbid bronchiectasis: A systemic review and meta-analysis. *Int. J. Chron. Obstruct. Pulmon. Dis.* **2015**, *10*, 1465–1475. [CrossRef]
45. Gao, Y.H.; Guan, W.J.; Liu, Q.; Wang, H.; Zhu, Y.; Chen, R.; Zhang, G. Impact of COPD and emphysema on survival of patients with lung cancer: A meta-analysis of observational studies. *Respirology* **2016**, *21*, 269–279. [CrossRef]
46. Song, Y.; Liu, J.; Lei, M.; Wang, Y.; Fu, Q.; Wang, B.; Guo, Y.; Mi, W.; Tong, L. An External-Validated Algorithm to Predict Postoperative Pneumonia Among Elderly Patients With Lung Cancer after Video-Assisted Thoracoscopic Surgery. *Front. Oncol.* **2021**, *11*, 777564. [CrossRef]
47. Sah, P.; Fitzpatrick, M.C.; Zimmer, C.F.; Abdollahi, E.; Juden-Kelly, L.; Moghadas, S.M.; Singer, B.H.; Galvani, A.P. Asymptomatic SARS-CoV-2 infection: A systematic review and meta-analysis. *Proc. Natl. Acad. Sci. USA* **2021**, *118*, e2109229118. [CrossRef]
48. Gattinoni, L.; Gattarello, S.; Steinberg, I.; Busana, M.; Palermo, P.; Lazzari, S.; Romitti, F.; Quintel, M.; Meissner, K.; Marini, J.J.; et al. COVID-19 pneumonia: Pathophysiology and management. *Eur. Respir. Rev.* **2021**, *30*, 210138. [CrossRef]
49. Bolatkale, M.; Duger, M.; Ülfer, G.; Can, Ç.; Acara, A.C.; Yiğitbaşı, T.; Seyhan, E.C.; Bulut, M. A novel biochemical marker for community-acquired pneumonia: Ischemia-modified albumin. *Am. J. Emerg. Med.* **2017**, *35*, 1121–1125. [CrossRef]
50. Chalmers, J.D.; Singanayagam, A.; Hill, A.T. C-reactive protein is an independent predictor of severity in community-acquired pneumonia. *Am. J. Med.* **2008**, *121*, 219–225. [CrossRef]
51. Fukuda, H.; Onizuka, H.; Nishimura, N.; Kiyohara, K. Risk factors for pneumococcal disease in persons with chronic medical conditions: Results from the LIFE Study. *Int. J. Infect. Dis.* **2022**, *116*, 216–222. [CrossRef]
52. Park, H.Y.; Song, W.J.; Yoo, H.; Eom, J.S.; Jeong, B.-H.; Lee, H.Y.; Koh, W.-J.; Jeon, K. Chronic obstructive pulmonary disease severity is associated with severe pneumonia. *Ann. Thorac. Med.* **2015**, *10*, 105–111. [CrossRef] [PubMed]
53. Kurashima, K.; Takaku, Y.; Hoshi, T.; Kanauchi, T.; Nakamoto, K.; Takayanagi, N.; Yanagisawa, T.; Sugita, Y.; Kawabata, Y. Lobe-based computed tomography assessment of airway diameter, airway or vessel number, and emphysema extent in relation to the clinical outcomes of COPD. *Int. J. Chron. Obstruct. Pulmon. Dis.* **2015**, *10*, 1027–1033. [CrossRef] [PubMed]
54. Pavord, I.D.; Lettis, S.; Anzueto, A.; Barnes, N. Blood eosinophil count and pneumonia risk in patients with chronic obstructive pulmonary disease: A patient-level meta-analysis. *Lancet Respir. Med.* **2016**, *4*, 731–741. [CrossRef] [PubMed]
55. Lee, A.L.; Goldstein, R.S. Gastroesophageal reflux disease in COPD: Links and risks. *Int. J. Chron. Obstruct. Pulmon. Dis.* **2015**, *10*, 1935–1949. [CrossRef] [PubMed]
56. Seemungal, T.A.; Donaldson, G.C.; Paul, E.A.; Bestall, J.C.; Jeffries, D.J.; Wedzicha, J.A. Effect of exacerbation on quality of life in patients with chronic obstructive pulmonary disease. *Am. J. Respir. Crit. Care Med.* **1998**, *157 Pt 1*, 1418–1422. [CrossRef]

57. Hsu, W.T.; Lai, C.C.; Wang, Y.H.; Tseng, P.-H.; Wang, K.; Wang, C.-Y.; Chen, L. Risk of pneumonia in patients with gastroesophageal reflux disease: A population-based cohort study. *PLoS ONE* **2017**, *12*, e0183808. [CrossRef]
58. Torres, A.; Niederman, M.S.; Chastre, J.; Ewig, S.; Fernandez-Vandellos, P.; Hanberger, H.; Kollef, M.; Bassi, G.L.; Luna, C.M.; Martin-Loeches, I.; et al. International ERS/ESICM/ESCMID/ALAT guidelines for the management of hospital-acquired pneumonia and ventilator-associated pneumonia: Guidelines for the management of hospital-acquired pneumonia (HAP)/ventilator-associated pneumonia (VAP) of the European Respiratory Society (ERS), European Society of Intensive Care Medicine (ESICM), European Society of Clinical Microbiology and Infectious Diseases (ESCMID) and Asociación Latinoamericana del Tórax (ALAT). *Eur. Respir. J.* **2017**, *50*, 1700582. [CrossRef]
59. Eber, M.R.; Laxminarayan, R.; Perencevich, E.N.; Malani, A. Clinical and economic outcomes attributable to health care-associated sepsis and pneumonia. *Arch. Intern. Med.* **2010**, *170*, 347–353. [CrossRef]
60. Sopena, N.; Sabrià, M.; Neunos 2000 Study Group. Multicenter study of hospital-acquired pneumonia in non-ICU patients. *Chest* **2005**, *127*, 213–219. [CrossRef]
61. Leu, H.S.; Kaiser, D.L.; Mori, M.; Woolson, R.F.; Wenzel, R.P. Hospital-acquired pneumonia. Attributable mortality and morbidity. *Am. J. Epidemiol.* **1989**, *129*, 1258–1267. [CrossRef]
62. Beach, T.G.; Russell, A.; Sue, L.I.; Intorcia, A.J.; Glass, M.J.; Walker, J.E.; Arce, R.; Nelson, C.M.; Hidalgo, T.; Chiarolanza, G.; et al. Increased Risk of Autopsy-Proven Pneumonia with Sex, Season and Neurodegenerative Disease. *medRxiv* **2021**, Preprint. [CrossRef]
63. Suetani, S.; Honarparvar, F.; Siskind, D.; Hindley, G.; Veronese, N.; Vancampfort, D.; Allen, L.; Solmi, M.; Lally, J.; Gaughran, F.; et al. Increased rates of respiratory disease in schizophrenia: A systematic review and meta-analysis including 619,214 individuals with schizophrenia and 52,159,551 controls. *Schizophr. Res.* **2021**, *237*, 131–140. [CrossRef] [PubMed]
64. Szylińska, A.; Bott-Olejnik, M.; Wańkowicz, P.; Karoń, D.; Rotter, I.; Kotfis, K. A Novel Index in the Prediction of Pneumonia Following Acute Ischemic Stroke. *Int. J. Environ. Res. Public Health* **2022**, *19*, 15306. [CrossRef]
65. Wei, M.; Zhao, Y.; Qian, Z.; Yang, B.; Xi, J.; Wei, J.; Tang, B. Pneumonia caused by Mycobacterium tuberculosis. *Microbes Infect.* **2020**, *22*, 278–284. [CrossRef]
66. Yakar, H.I.; Gunen, H.; Pehlivan, E.; Aydogan, S. The role of tuberculosis in COPD. *Int. J. Chron. Obstruct. Pulmon. Dis.* **2017**, *12*, 323–329. [CrossRef]
67. Hirano, R.; Fujita, M.; Matsumoto, T.; On, R.; Watanabe, K. Inhaled corticosteroids might not increase the risk of pneumonia in patients with chronic obstructive pulmonary disease in Japan. *Int. J. Chron. Obstruct. Pulmon. Dis.* **2018**, *13*, 3503–3509. [CrossRef]
68. Iannella, H.A.; Luna, C.M. Community-Acquired Pneumonia in Latin America. *Semin. Respir. Crit. Care Med.* **2016**, *37*, 868–875. [CrossRef] [PubMed]
69. McQuitty, R.; Bui, R.; Chaaban, M.R. Retrospective Study: Association of Chronic Sphenoid Rhinosinusitis With Community Acquired Pneumonia. *Am. J. Rhinol. Allergy* **2019**, *33*, 751–756. [CrossRef]
70. Williams, N.P.; Coombs, N.A.; Johnson, M.J.; Josephs, L.; A Rigge, L.; Staples, K.; Thomas, M.; Wilkinson, T.M. Seasonality, risk factors and burden of community-acquired pneumonia in COPD patients: A population database study using linked health care records. *Int. J. Chron. Obstruct. Pulmon. Dis.* **2017**, *12*, 313–322. [CrossRef]
71. Juthani-Mehta, M.; De Rekeneire, N.; Allore, H.; Chen, S.; Ma, J.R.O.; Bauer, D.C.; Harris, T.B.; Newman, A.B.; Yende, S.; Weyant, R.; et al. Modifiable risk factors for pneumonia requiring hospitalization of community-dwelling older adults: The Health, Aging, and Body Composition Study. *J. Am. Geriatr. Soc.* **2013**, *61*, 1111–1118. [CrossRef]
72. Halpin, D.M.G.; Criner, G.J.; Dransfield, M.T.; Han, M.K.; Hartley, B.; Harvey, C.; Jones, C.E.; Kato, M.; Lange, P.; Lettis, S.; et al. Triple Versus Dual Combination Therapy in Chronic Obstructive Pulmonary Disease in Asian Countries: Analysis of the IMPACT Trial. *Pulm. Ther.* **2021**, *7*, 101–118. [CrossRef] [PubMed]
73. Kurashima, K.; Takaku, Y.; Nakamoto, K.; Kanauchi, T.; Takayanagi, N.; Yanagisawa, T.; Sugita, Y.; Araki, R. Risk Factors for Pneumonia and the Effect of the Pneumococcal Vaccine in Patients With Chronic Airflow Obstruction. *Chronic Obstr. Pulm. Dis.* **2016**, *3*, 610–619. [CrossRef] [PubMed]
74. Make, B.; Dutro, M.P.; Paulose-Ram, R.; Marton, J.P.; Mapel, D.W. Undertreatment of COPD: A retrospective analysis of US managed care and Medicare patients. *Int. J. Chron. Obstruct. Pulmon. Dis.* **2012**, *7*, 1–9. [CrossRef] [PubMed]
75. Di Pasquale, M.F.; Sotgiu, G.; Gramegna, A.; Radovanovic, D.; Terraneo, S.; Reyes, L.F.; Rupp, J.; Del Castillo, J.G.; Blasi, F.; Aliberti, S.; et al. Prevalence and Etiology of Community-acquired Pneumonia in Immunocompromised Patients. *Clin. Infect. Dis.* **2019**, *68*, 1482–1493. [CrossRef]
76. Aliberti, S.; Cilloniz, C.; Chalmers, J.D.; Zanaboni, A.M.; Cosentini, R.; Tarsia, P.; Pesci, A.; Blasi, F.; Torres, A. Multidrug-resistant pathogens in hospitalised patients coming from the community with pneumonia: A European perspective. *Thorax* **2013**, *68*, 997–999. [CrossRef]
77. Bhat, T.A.; Panzica, L.; Kalathil, S.G.; Thanavala, Y. Immune Dysfunction in Patients with Chronic Obstructive Pulmonary Disease. *Ann. Am. Thorac. Soc.* **2015**, *12* (Suppl. S2), S169–S175. [CrossRef]
78. Huang, L.; Crothers, K. HIV-associated opportunistic pneumonias. *Respirology* **2009**, *14*, 474–485. [CrossRef]
79. Zifodya, J.S.; Triplette, M.; Shahrir, S.; Attia, E.F.M.; Akgun, K.M.M.; Hoo, G.W.S.; Rodriguez-Barradas, M.C.; Wongtrakool, C.; Huang, L.; Crothers, K. A cross-sectional analysis of diagnosis and management of chronic obstructive pulmonary disease in people living with HIV: Opportunities for improvement. *Medicine* **2021**, *100*, e27124. [CrossRef]

80. Antoniou, T.; Yao, Z.; Raboud, J.; Gershon, A.S. Incidence of chronic obstructive pulmonary disease in people with HIV in Ontario, 1996-2015: A retrospective population-based cohort study. *CMAJ Open* **2020**, *8*, E83–E89. [CrossRef]
81. Raynaud, C.; Roche, N.; Chouaid, C. Interactions between HIV infection and chronic obstructive pulmonary disease: Clinical and epidemiological aspects. *Respir. Res.* **2011**, *12*, 117. [CrossRef]
82. Pascoe, S.J.; Papi, A.; Midwinter, D.; Lettis, S.; Barnes, N. Circulating neutrophils levels are a predictor of pneumonia risk in chronic obstructive pulmonary disease. *Respir. Res.* **2019**, *20*, 195. [CrossRef]
83. de Jager, C.P.; Wever, P.C.; Gemen, E.F.; Kusters, R.; Van Gageldonk-Lafeber, A.B.; Van Der Poll, T.; Laheij, R.J.F. The neutrophil-lymphocyte count ratio in patients with community-acquired pneumonia. *PLoS ONE* **2012**, *7*, e46561. [CrossRef]
84. Lee, H.; Kim, I.; Kang, B.H.; Um, S.J. Prognostic value of serial neutrophil-to-lymphocyte ratio measurements in hospitalized community-acquired pneumonia. *PLoS ONE* **2021**, *16*, e0250067. [CrossRef]
85. Paliogiannis, P.; Fois, A.G.; Sotgia, S.; Mangoni, A.A.; Zinellu, E.; Pirina, P.; Negri, S.; Carru, C.; Zinellu, A. Neutrophil to lymphocyte ratio and clinical outcomes in COPD: Recent evidence and future perspectives. *Eur. Respir. Rev.* **2018**, *27*, 170113. [CrossRef] [PubMed]
86. Hsu, S.C.; Chang, J.H.; Hsu, Y.P.; Bai, K.J.; Huang, S.K.; Hsu, C.W. Circulating sphingosine-1-phosphate as a prognostic biomarker for community-acquired pneumonia. *PLoS ONE* **2019**, *14*, e0216963. [CrossRef] [PubMed]
87. Quraishi, S.A.; Bittner, E.A.; Christopher, K.B.; Camargo, C.A., Jr. Vitamin D status and community-acquired pneumonia: Results from the third National Health and Nutrition Examination Survey. *PLoS ONE* **2013**, *8*, e81120. [CrossRef] [PubMed]
88. Fine, M.J.; Auble, T.E.; Yealy, D.M.; Hanusa, B.H.; Weissfeld, L.A.; Singer, D.E.; Coley, C.M.; Marrie, T.J.; Kapoor, W.N. A prediction rule to identify low-risk patients with community-acquired pneumonia. *N. Engl. J. Med.* **1997**, *336*, 243–250. [CrossRef]
89. Ferguson, G.T.; Rabe, K.F.; Martinez, F.J.; Fabbri, L.M.; Wang, C.; Ichinose, M.; Bourne, E.; Ballal, S.; Darken, P.; DeAngelis, K.; et al. Triple therapy with budesonide/glycopyrrolate/formoterol fumarate with co-suspension delivery technology versus dual therapies in chronic obstructive pulmonary disease (KRONOS): A double-blind, parallel-group, multicentre, phase 3 randomised controlled trial. *Lancet Respir. Med.* **2018**, *6*, 747–758. [CrossRef]
90. Wedzicha, J.A.; Banerji, D.; Chapman, K.R.; Vestbo, J.; Roche, N.; Ayers, R.T.; Thach, C.; Fogel, R.; Patalano, F.; Vogelmeier, C.F. Indacaterol-Glycopyrronium versus Salmeterol-Fluticasone for COPD. *N. Engl. J. Med.* **2016**, *374*, 2222–2234. [CrossRef]
91. Bardsley, S.; Criner, G.J.; Halpin, D.M.G.; Han, M.K.; Hanania, N.A.; Hill, D.; Lange, P.; Lipson, D.A.; Martinez, F.J.; Midwinter, D.; et al. Single-inhaler triple therapy fluticasone furoate/umeclidinium/vilanterol versus dual therapy in current and former smokers with COPD: IMPACT trial post hoc analysis. *Respir. Med.* **2022**, *205*, 107040. [CrossRef]
92. Cazzola, M.; Rogliani, P.; Laitano, R.; Calzetta, L.; Matera, M.G. Beyond Dual Bronchodilation—Triple Therapy, When and Why. *Int. J. Chron. Obstruct. Pulmon. Dis.* **2022**, *17*, 165–180. [CrossRef] [PubMed]

MDPI AG
Grosspeteranlage 5
4052 Basel
Switzerland
Tel.: +41 61 683 77 34

Biomedicines Editorial Office
E-mail: biomedicines@mdpi.com
www.mdpi.com/journal/biomedicines

www.ingramcontent.com/pod-product-compliance
Lightning Source LLC
Chambersburg PA
CBHW062057220926
51850CB00010B/2800
9783725855551